*LECTURES ON THE THEORY OF
INTEGRAL
EQUATIONS*

OTHER *GRAYLOCK* PUBLICATIONS

KHINCHIN: *Three Pearls of Number Theory*
PONTRYAGIN: *Foundations of Combinatorial Topology*
NOVOZHILOV: *Foundations of the Nonlinear Theory of Elasticity*
ALEKSANDROV: *Combinatorial Topology, Vol. 1*
KOLMOGOROV and FOMIN: *Elements of the Theory of Functions and Functional Analysis, Vol. 1. Metric and Normed Spaces*

LECTURES ON THE THEORY OF
INTEGRAL EQUATIONS

BY

I. G. PETROVSKIĬ

TRANSLATED FROM THE SECOND REVISED (1951) RUSSIAN EDITION

by

HYMAN KAMEL AND HORACE KOMM

Department of Mathematics
Rensselaer Polytechnic Institute

GRAYLOCK PRESS

ROCHESTER, N. Y.

1957

Copyright, 1957, by
GRAYLOCK PRESS
Rochester, N. Y.

All rights reserved. This book, or parts thereof, may not be reproduced in any form, or translated, without permission in writing from the publishers.

Manufactured in the United States of America

CONTENTS

Chapter I
INTRODUCTION. THE FREDHOLM THEOREMS

1. Definitions. Examples.. 1
2. Typical problems leading to integral equations................. 2
3. Analogy between linear integral equations and linear algebraic equations. Formulation of the Fredholm theorems............ 6
4. Integral equations with degenerate kernels.................... 10
5. Integral equations with continuous kernels of sufficiently small absolute bound... 17
6. Integral equations with almost degenerate kernels............. 24
7. Integral equations with uniformly continuous kernels.......... 28
8. Integral equations with kernels of the form $\bar{K}(P,Q)/PQ^\alpha$........ 29
9. Examples of singular integral equations...................... 38

Chapter II
VOLTERRA INTEGRAL EQUATIONS

10. Volterra integral equations................................. 40

Chapter III
INTEGRAL EQUATIONS WITH REAL SYMMETRIC KERNELS

11. Geometric analogies to certain relations between functions (function space)... 44
12. The proof of the existence of eigenfunctions for integral equations with symmetric kernels...................................... 54
13. Some properties of eigenfunctions and eigenvalues for integral equations with symmetric kernels............................ 61
14. The Hilbert-Schmidt theorem................................ 68
15. A theorem on the expansion of the kernel.................... 72
16. Classification of kernels.................................... 73
17. Dini's theorem and its applications........................... 74
18. Example... 77

APPENDIX

19. Reduction of a quadratic form to canonical form by means of an orthogonal transformation................................. 80

20. Theory of integral equations with symmetric kernels that are Lebesgue square integrable............................... 85
List of theorems.. 95
Index... 97

Chapter I
INTRODUCTION
THE FREDHOLM THEOREMS

§1. Definitions. Examples

Integral equations are equations that contain the unknown function under the integral sign. In particular, the equation

$$(1.1) \qquad a(x)\varphi(x) + f(x) = \int_a^b K(x, \xi)\varphi(\xi)\, d\xi$$

is an integral equation in the function $\varphi(\xi)$, where $a(x)$, $f(x)$, $K(x, \xi)$ are known functions and $\varphi(\xi)$ is to be determined. The variables x and ξ take on all values of the interval (a, b).

In this book we shall consider only equations which contain the unknown function linearly, i.e. only equations of type (1.1). Such equations are called *linear integral equations*. If $a(x)$ never vanishes, division of both sides of (1.1) by $a(x)$ yields an equation of the form

$$(1.2) \qquad \varphi(x) = \int_a^b K(x, \xi)\varphi(\xi)\, d\xi + f(x).$$

Such equations are called linear integral equations of the *second kind* or *Fredholm* integral equations after the mathematician who first investigated them. If $f(x) \equiv 0$, then equation (1.2) is said to be *homogeneous*.

If $a(x) \equiv 0$, equation (1.1) becomes

$$\int_a^b K(x, \xi)\varphi(\xi)\, d\xi = f(x).$$

It is called a linear integral equation of the *first kind*.

The function $K(x, \xi)$ is referred to as the *kernel* of the integral equation.

In the sequel we shall deal primarily with linear integral equations of the second kind.

One could also consider integral equations in which the unknown function is dependent not only on one variable but on several. Such, for example, is the equation

$$\varphi(x, y) = \int_G K(x, y; \xi, \eta)\varphi(\xi, \eta)\, d\xi\, d\eta + f(x, y)$$

in the unknown function $\varphi(x, y)$, where G is a region of the (ξ, η)-plane, and (x, y) is a point of G. Such an equation may be written in the form

$$\varphi(P) = \int_G K(P, Q)\varphi(Q)\, dQ + f(P),$$

where $P \in G$ and $Q \in G$. ($A \in M$ signifies that the point A is contained in the set M.)

One could also consider systems of integral equations with several unknown functions.

REMARK. Except for §20, we shall always assume, even though this may not be explicitly stated, that the functions of the point P or Q are defined in a bounded d-dimensional domain G, and that they are continuous everywhere in G except possibly at a finite number of points, sufficiently smooth curves, and surfaces of dimension $\leq d - 1$. The functions need not be defined on these exceptional points, curves, and surfaces. The boundary of the region G is to consist of a finite number of pieces of smooth $(d - 1)$-dimensional surfaces, or, if $d = 2$, of a finite number of smooth arcs.

With the exception of §20, all integrations are to be understood in the ordinary sense provided the functions are continuous in G. If the functions have discontinuities at certain points, curves, or surfaces, the integrals are to be taken as improper. We shall assume that all functions to be considered are absolutely integrable.

§2. Typical problems leading to integral equations

Consider an elastic string of length l that is perfectly flexible. An increase in length of Δl requires (Hooke's law) a force equal to $c\Delta l$, where c is a constant. The ends of the string are fixed at the points A and B on the nonnegative x-axis, A coincides with the origin of the horizontal x-axis. If the string, in rest position, is subjected only to a horizontal stretching force T_0, the string will be horizontal, i.e., it will coincide with the Ox-axis. T_0 is to be very large in comparison with all other forces considered.

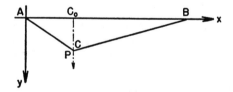

FIG. 1

At the point C with abscissa $x = \xi$, a vertical force P is applied. Under the influence of P the string assumes the form of the broken line ACB (Fig. 1). Since P is to be small in comparison with T_0, $C_0C = \delta$ will be small in comparison with AC_0 and C_0B. Neglecting δ^2 in comparison with l, we may assume that the tension in the string remains equal to T_0 even under the action of the force P. Taking into account the vertical components of the tension forces in the string at C and the force P, again neglecting terms in δ^2, we obtain

$$T_0\delta/\xi + T_0\delta/(l - \xi) = P,$$

whence $\delta = P(l - \xi)\xi/T_0 l$. If one denotes by $y(x)$ the deflection of the string at the point with abscissa x, then one has

$$y(x) = PG(x, \xi),$$

where

(2.1) $\begin{cases} G(x, \xi) = x(l - \xi)/T_0 l \text{ for the interval } AC (0 \leq x \leq \xi). \\ G(x, \xi) = (l - x)\xi/T_0 l \text{ for the interval } CB (\xi \leq x \leq l). \end{cases}$

Making use of these formulae it is easily verified that

$$G(x, \xi) = G(\xi, x).$$

We now assume that the string is subjected to a continuously distributed vertical force with linear density $p(\xi)$ so that the force acting on the piece between ξ and $\xi + \Delta\xi$ is approximately $p(\xi)\Delta\xi$. Since one may sum the displacements due to the forces $p(\xi)\Delta\xi$ ("Principle of Superposition"), the string assumes the form

$$y(x) = \int_0^l G(x, \xi) p(\xi)\, d\xi.$$

We shall now treat the following problems.

1. Find the force density distribution $p(\xi)$ under whose influence the string assumes a given form $y = y(x)$. This leads to an integral equation of the first kind

(2.2) $$y(x) = \int_0^l G(x, \xi) p(\xi)\, d\xi$$

for the function $p(\xi)$.

2. Let the force density distribution be

$$p(\xi) \sin \omega t \qquad\qquad (\omega > 0),$$

where t is the time. The string is set in motion under its influence. We assume that the abscissa of no moving point is altered and that the string undergoes a periodic oscillation given by

$$y = y(x) \sin \omega t.$$

If we designate the linear mass density of the string by $\rho(\xi)$ at the point ξ, then it follows that at time t the inertial force

$$-\rho(\xi)\Delta\xi\, d^2y/dt^2 = \rho(\xi)y(\xi)\omega^2 \sin \omega t\, \Delta\xi$$

as well as the force $p(\xi)\Delta\xi$ operate on the section of string between ξ and $\xi + \Delta\xi$. Equation (2.2) then assumes the following form:

$$y(x) \sin \omega t = \int_0^l G(x, \xi)[p(\xi) \sin \omega t + \omega^2 \rho(\xi) y(\xi) \sin \omega t]\, d\xi.$$

Dividing through by $\sin \omega t$ and making the substitutions

$$\int_0^l G(x, \xi)p(\xi)\, d\xi = f(x), \qquad G(x, \xi)\rho(\xi) = K(x, \xi), \qquad \omega^2 = \lambda,$$

we get

(2.3) $$y(x) = \lambda \int_0^l K(x, \xi) y(\xi)\, d\xi + f(x).$$

If $p(\xi)$ is given and thereby also $f(x)$, we have a Fredholm integral equation for the determination of the function $y(x)$. From the definition of the function $f(x)$ it follows that

$$f(0) = f(l) = 0.$$

If the density $\rho(\xi)$ is constant and $f(x)$ is a twice continuously differentiable function, then the integral equation is easily solvable. In fact, if we substitute the expression (2.1) for $G(x, \xi)$ into $K(x, \xi)$, we obtain

$$y(x) = \omega^2 \rho \int_0^x [(l - x)\xi/(T_0 l)] y(\xi)\, d\xi$$
$$+ \omega^2 \rho \int_x^l [x(l - \xi)/(T_0 l)] y(\xi)\, d\xi + f(x)$$

or

$$y(x) = [\omega^2 c(l - x)/l] \int_0^x \xi y(\xi)\, d\xi + (\omega^2 c x/l) \int_x^l (l - \xi) y(\xi)\, d\xi + f(x),$$

where

$$c = \rho/T_0.$$

Differentiating twice with respect to x one obtains the equation

(2.4) $$y''(x) = -\omega^2 c y(x) + f''(x).$$

On the other hand, one may show that every solution of the differential equation (2.4) which vanishes for $x = 0$ and $x = l$ is also a solution of the integral equation (2.3). To this end we multiply both sides of the equation

$$y''(\xi) = -\omega^2 c y(\xi) + f''(\xi)$$

by $-T_0 G(x, \xi)$. After integrating with respect to ξ from 0 to l one obtains the equation (2.3) since

$$\int_0^l T_0 G(x, \xi) \varphi''(\xi) \, d\xi = -\varphi(x);$$

this relation follows easily by two integrations by parts. $\varphi(x)$ is required to be an arbitrary twice continuously differentiable function vanishing at $x = 0$ and $x = l$.

As is known from the theory of ordinary differential equations, the general solution of the equation (2.4) is

$$y = c_1 \sin \mu x + c_2 \cos \mu x + (1/\mu) \int_0^x f''(\xi) \sin \mu(x - \xi) \, d\xi,$$

$\mu = \omega c^{\frac{1}{2}}$ and c_1, c_2 are arbitrary constants. From the equations (2.1) it follows that $y(0) = y(l) = 0$. Determining the constants of integration from these conditions we get

(2.5)
$$y(x) = -[\sin \mu x/(\mu \sin \mu l)] \int_0^l f''(\xi) \sin \mu(l - \xi) \, d\xi$$
$$+ (1/\mu) \int_0^x f''(\xi) \sin \mu(x - \xi) \, d\xi$$

if $\sin \mu l \neq 0$.

In this case the integral equation (2.3) has a unique solution for an arbitrary function $f(x)$, provided only that the latter is twice continuously differentiable and $f(0) = f(l) = 0$.

It can be shown that the continuity of $f(x)$ is sufficient to insure the existence of a solution of (2.3) provided $\sin \mu l \neq 0$. The requirement of the existence, let alone the continuity of the second derivative of $f(x)$ is superfluous. But the requirement that $\sin \mu l \neq 0$ is necessary in order that the integral equation have a solution for every continuous function or even for every sufficiently often differentiable function $f(x)$.

If $\sin \mu l = 0$, it follows that

(2.6) $$\mu = k\pi/l,$$

(2.7) $$\omega = k\pi/lc^{\frac{1}{2}},$$

(2.8) $$\lambda = k^2\pi^2/l^2 c,$$

where k is an arbitrary integer positive, negative, or zero. The values of the parameter λ given by (2.8) for $k = 1, 2, 3, \cdots$ are called the *eigenvalues* of the integral equation (2.3), the corresponding values of ω the *eigenfrequencies* of the vibrating string.

It follows from the derivation of formula (2.5) that the integral equation (2.3) has a solution in case $\sin \mu l = 0$ and $f(x)$ has a continuous second derivative, provided $f(x)$ satisfies

$$(2.9) \qquad \int_0^l f''(\xi) \sin \mu(l - \xi) \, d\xi = 0.$$

Integrating by parts and making use of the facts that $\sin \mu(l - \xi)$ and $f(\xi) = 0$ for $\xi = 0$ and $\xi = l$, one may write the last condition in the form

$$(2.10) \qquad \int_0^l f(\xi) \sin \mu \xi \, d\xi = 0.$$

Conversely, one may easily convince oneself that condition (2.10) for a given μ is also sufficient for the existence of a solution of equation (2.3) provided $\sin \mu l = 0$.

The condition (2.10) is satisfied in particular if

$$f(x) \equiv 0.$$

The integral equation (2.3) and the differential equation (2.4) are then homogeneous. All solutions of the homogeneous differential equation (2.4) that vanish for $x = 0$ and $x = l$ and consequently all solutions of the integral equation (2.3) have the form

$$(2.11) \qquad y(x) = C \sin \mu_k x,$$

where C is an arbitrary constant and μ_k is one of the numbers given by (2.6). The formula yields the amplitudes of the eigenoscillations of the string at the point x:

$$y = C \sin \mu_k x \sin \omega_k t,$$

which arise in the absence of any external force. Evidently such oscillations cannot have arbitrary frequencies but only such as are given by (2.7) for $k = 1, 2, 3, \cdots$.

If condition (2.9) is not satisfied, then formula (2.5) shows that the amplitude $y(x)$ of the periodic oscillation becomes infinitely large as the frequency of oscillation of the external force approaches one of the eigenfrequencies of the string. At the coincidence of these frequencies resonance takes place. In this case there do not exist periodic oscillations in general, that is for arbitrary amplitudes of the external force.

Correspondingly, there exists in general no solution of the non-homogeneous integral equation (2.3) for λ equal to an eigenvalue of this equation.

§3. Analogy between linear integral equations and linear algebraic equations. Formulation of the Fredholm theorems

Let us consider a linear integral equation of the second kind

$$(3.1) \qquad y(x) = \int_a^b K(x, \xi) y(\xi) \, d\xi + f(x),$$

where $K(x, \xi)$ and $f(x)$ are known functions for $a \leq x \leq b$ and $a \leq \xi \leq b$. Divide the interval (a, b) into n equal parts of length

$$(b - a)/n = \Delta x = \Delta \xi.$$

Let us put

$$K(a + p\Delta x, a + q\Delta \xi) = K_{pq} \quad (p, q = 1, 2, \cdots, n),$$

$$y(a + p\Delta x) = y_p \quad (p = 1, 2, \cdots, n),$$

$$f(a + p\Delta x) = f_p \quad (p = 1, 2, \cdots, n).$$

We now replace the integral $\int_a^b K(x, \xi) y(\xi) \, d\xi$ for $x = a + p\Delta x$ by the sum

$$\sum_{q=1}^n K_{pq} y_q \Delta \xi, \qquad p = 1, 2, \cdots, n.$$

We then have in place of the integral equation (3.1) the system of linear algebraic equations

(3.2) $$y_p = \sum_{q=1}^n K_{pq} y_q \Delta \xi + f_p, \qquad p = 1, 2, \cdots, n.$$

The K_{pq}, f_p, and $\Delta \xi$ are known and the y_p are to be found.

The purpose of the following paragraphs is to carry over the known theorems about linear algebraic equations to the Fredholm integral equations of the second kind. Determinants are involved in the usual formulation of the theorems on linear algebraic equations. While it is possible to carry determinants over to integral equations, it is inconvenient. Hence we shall formulate these theorems without determinants. We express them here recursively.

The coefficient determinant of the system

(3.3)
$$\begin{vmatrix} 1 - K_{11}\Delta\xi & -K_{12}\Delta\xi & \cdots & -K_{1n}\Delta\xi \\ -K_{21}\Delta\xi & 1 - K_{22}\Delta\xi & \cdots & -K_{2n}\Delta\xi \\ \cdots & \cdots & \cdots & \cdots \\ -K_{n1}\Delta\xi & -K_{n2}\Delta\xi & \cdots & 1 - K_{nn}\Delta\xi \end{vmatrix}$$

is essential for the solution of the system (3.2). If this determinant is not zero, it is known that the system (3.2) always has a unique solution for arbitrary values of $f_1, f_2, \cdots, f_n$. In this case the transposed system

$$z_p = \sum_{q=1}^n K_{qp} z_q \Delta \xi + f_p^*, \qquad p = 1, 2, \cdots, n,$$

also has a unique solution for arbitrary f_p^*.

If the determinant is zero, then the system (3.2) does not in general have a solution for arbitrary f_p. In this case however the corresponding homogeneous system, i.e. the system arising from (3.2) if all the f_p are put equal to zero, always has a non-trivial solution, that is, a solution which does not consist of all zeros.

Thus the following alternatives hold. *Either the given non-homogeneous system of linear algebraic equations* (3.2) *has one and only one solution for arbitrary* $f_1, f_2, \cdots, f_n$ *or the corresponding homogeneous system has at least one non-trivial solution. In the first case the same holds for the transposed system.*

In the second case, the homogeneous system

$$(3.4) \qquad y_p - \sum_{q=1}^{n} K_{pq} y_q \Delta \xi = 0, \qquad p = 1, \cdots, n,$$

has the same number of linearly independent solutions as its transposed system

$$(3.5) \qquad z_p - \sum_{q=1}^{n} K_{qp} z_q \Delta \xi = 0, \qquad p = 1, \cdots, n.$$

This number is $n - r$ *if the rank of the matrix of the determinant* (3.3) *is* r.

The assertion of the existence of exactly $n - r$ linearly independent solutions of the homogeneous systems (3.4) and (3.5) is valid also for the first case if $n = r$. The expression "zero linearly independent solutions" signifies that there is precisely one solution consisting only of zeros.

We shall now find necessary and sufficient conditions for the non-homogeneous system (3.2) to have a solution in the second case. It is easier to find a necessary condition. Suppose $z_1, z_2, \cdots, z_n$ is any solution of (3.5). Multiply the pth of equations (3.2) by z_p and add all the equations to obtain

$$\sum_p y_p z_p - \sum_{p,q} K_{pq} y_q z_p \Delta \xi = \sum_p f_p z_p.$$

One may write the left side in the form

$$\sum_p y_p z_p - \sum_{p,q} K_{pq} y_p z_q \Delta \xi = \sum_p y_p (z_p - \sum_q K_{qp} z_q \Delta \xi).$$

In virtue of (3.5) this expression is equal to zero, and one must therefore have

$$(3.6) \qquad \sum_{p=1}^{n} f_p z_p = 0.$$

We now show that this condition is also sufficient for the existence of a solution of the system (3.2), provided it is satisfied for every solution of the system (3.5). Obviously this last requirement will be met if (3.6) is satisfied for any $(n - r)$ linearly independent solutions of the system (3.5). It is known from algebra that if the determinant vanishes, it is sufficient for the existence of a solution of the system (3.2) that the rank of the matrix

(3.7)
$$\begin{Vmatrix} 1 - K_{11}\Delta\xi & -K_{12}\Delta\xi & \cdots & -K_{1n}\Delta\xi & f_1 \\ -K_{21}\Delta\xi & 1 - K_{22}\Delta\xi & \cdots & -K_{2n}\Delta\xi & f_2 \\ \cdot\cdot\cdot\cdot\cdot\cdot\cdot\cdot\cdot\cdot\cdot\cdot\cdot\cdot\cdot\cdot\cdot\cdot\cdot\cdot\cdot\cdot\cdot\cdot \\ -K_{n1}\Delta\xi & -K_{n2}\Delta\xi & \cdots & 1 - K_{nn}\Delta\xi & f_n \end{Vmatrix}$$

be equal to the rank of the matrix of (3.3).

For equality it is sufficient that every $(r + 1)$st order determinant formed from (3.7) and containing elements of the last column be equal to zero. If such a determinant D_{r+1} is developed according to the elements f_k, then condition (3.6) will insure equality to zero. To show this we shall construct numbers $z_1, z_2, \cdots, z_n$ as follows to satisfy the system (3.5): For an i for which f_i is in D_{r+1} choose z_i equal to the cofactor of f_i in D_{r+1}, otherwise set $z_i = 0$.

The correctness of the assertion may be shown as follows. The numbers $z_1, z_2, \cdots, z_n$ are inserted in the jth equation of system (3.5). If the elements in the jth column of matrix (3.7) are contained in D_{r+1}, then the result of the substitution is zero, since we get a determinant with two equal columns. If the elements in the jth column are not contained in the determinant D_{r+1}, it equals zero anyway, since we have a determinant of order $r + 1$ from the matrix which is of rank r.

Thus the following holds: *In the second case there is a solution of the non-homogeneous system if, and only if, condition* (3.6) *is satisfied for an arbitrary solution* $(z_1, z_2, \cdots, z_n)$ *of the transposed homogeneous system.*

We remark further that in this case the solution of the system (3.2) is not unique. If one adds to a solution of (3.2) an arbitrary solution of the corresponding homogeneous system, one again obtains a solution of (3.2).

As $\Delta\xi$ tends to zero it is natural to expect that $\sum_q K_{pq}y_q\Delta\xi$ goes over into the integral $\int_a^b K(x, \xi)y(\xi)\,d\xi$ and the solution of the system of equations (3.2) into the solution of the integral equation (3.1). This is indeed true under certain assumptions regarding the kernel $K(x, \xi)$. Since the proof of this theorem is very unwieldy it will be omitted here, although for an approximate solution the integral equation (3.1) is sometimes replaced by the system (3.2) [1][1]. Here it will be shown only that the theorems formulated above for the system (3.2) go over into the following:

THEOREM 1. *Either the given linear non-homogeneous integral equation of the second kind has one and only one solution for every function* $f(x)$ *(from some sufficiently large class), or the corresponding homogeneous equation has at least one non-trivial solution, i.e. one that is not identically zero.*

THEOREM 2. *If the first alternative occurs for the given equation* (3.1), *then*

---

[1] Numbers in brackets refer to the references cited at the end of the chapter.

this also holds for the transposed equation

$$z(x) = \int_a^b K(\xi, x)z(\xi) \, d\xi + f^*(x).$$

For either alternative, the corresponding homogeneous integral equation and its transpose have the same finite number of linearly independent solutions.

If the functions $y_1(x), y_2(x), \cdots, y_n(x)$ satisfy the homogeneous equation (3.1), then evidently an arbitrary linear combination $C_1 y_1(x) + C_2 y_2(x) + \cdots + C_n y_n(x)$ with constant coefficients C_i is a solution of this equation.

THEOREM 3. *In the second case a necessary and sufficient condition for the existence of a solution of the non-homogeneous equation* (3.1) *is that*

$$\int_a^b f(x)z(x) \, dx = 0,$$

where $z(x)$ is any solution of the transposed homogeneous equation of (3.1).

If this condition is satisfied, equation (3.1) has infinitely many solutions since, as one easily shows, all functions of the form

$$y(x) + \varphi(x)$$

satisfy the equation, where $y(x)$ is any solution of equation (3.1) and $\varphi(x)$ is an arbitrary solution of the corresponding homogeneous equation. On the other hand it is clear that the difference of two solutions of (3.1) satisfies the corresponding homogeneous equation.

Theorems 1–3 formulated above are called the Fredholm theorems since Fredholm demonstrated them for equation (3.1) under sufficiently strong assumptions about the functions $K(x, \xi)$ and $f(x)$. The following sections are devoted to a proof of these theorems for certain classes of equations. For this purpose the number of independent variables is unessential. Hence all proofs will be carried out for an arbitrary number of independent variables. Just as at the end of §1, we shall write P for x and Q for ξ. These proofs as do, in general, the majority of existence proofs also yield methods for the approximate solutions of the integral equation (3.1).

The first Fredholm theorem as a statement of exclusive alternatives is particularly important for applications. Instead of showing that the given integral equation (3.1) has a solution, it is often simpler to show that the corresponding homogeneous equation or its transposed equation has only trivial solutions. And from this it follows according to the first theorem that the given integral equation (3.1) has indeed a solution.

§4. Integral equations with degenerate kernels

There is a class of integral equations that is easily transformed to linear algebraic equations. The Fredholm theorems for these equations can be immediately derived from the theorems on linear algebraic equations formulated in the preceding section. This class consists of integral equations with so called *degenerate kernels*.

We now demonstrate the Fredholm theorems for integral equations with degenerate kernels and use this special case in the sequel to prove the Fredholm theorems for integral equations with arbitrary continuous kernels.

A kernel is called *degenerate* if it has the form

(4.1) $$K(P, Q) = \sum_{i=1}^{m} a_i(P) b_i(Q).$$

We shall assume that $a_i(P)$, $b_i(Q)$, $y(P)$ and $f(P)$ are uniformly continuous in a given bounded domain G and that the $a_i(P)$ as well as the $b_i(Q)$ form linearly independent sets.

Let us show that the last assumption causes no restriction of generality. For suppose that there were constants $C_1, \cdots, C_m$ for which

$$C_1 a_1(P) + \cdots + C_m a_m(P) \equiv 0,$$

with at least one of $C_1, \cdots, C_m$ not 0. Suppose $C_m \neq 0$. The equation can be solved for a_m, yielding

$$a_m(P) \equiv C_1^* a_1(P) + \cdots + C_{m-1}^* a_{m-1}(P).$$

Substituting this expression in the right side of (4.1), we obtain

$$K(P, Q) \equiv \sum_{i=1}^{m-1} a_i(P) b_i(Q) + \sum_{i=1}^{m-1} C_i^* a_i(P) b_m(Q)$$
$$\equiv \sum_{i=1}^{m-1} a_i(P) [b_i(Q) + C_i^* b_m(Q)] \equiv \sum_{i=1}^{m-1} a_i(P) b_i^*(Q).$$

Thus the kernel may be written as a sum of fewer than m products of functions of P by functions of Q. If now the functions $a_i(P)$ or $b_i^*(Q)$, $i = 1, \cdots, m - 1$, again are linearly dependent, the number may be reduced again, etc.

As has already been stated, integral equations with degenerate kernels are easily transformed into linear equations and for these the Fredholm theorems are easily proved. Suppose the integral equation

(4.2) $$y(P) = \int_G K(P, Q) y(Q) \, dQ + f(P),$$

for which $K(P, Q)$ has the form (4.1), has a solution. Then

$$y(P) = \int \sum_{i=1}^{m} a_i(P) b_i(Q) y(Q) \, dQ + f(P)$$

or

(4.3) $$y(P) = \sum_{i=1}^{m} a_i(P) \int b_i(Q) y(Q) \, dQ + f(P).$$

Here as everywhere in the following we omit the symbol G from the integral sign. The symbol $\int$ is to signify always an integral over the domain G.

If we put for brevity

(4.4) $$\int b_i(Q) y(Q)\, dQ = C_i,$$

then equation (4.3) becomes

(4.5) $$y(P) = \sum_i C_i a_i(P) + f(P).$$

To determine the constants C_i we put the above expression for $y(P)$ in (4.4). We obtain

$$\int b_i(Q)[\sum_j C_j a_j(Q) + f(Q)]\, dQ = C_i.$$

If one sets

(4.6) $$\int b_i(Q) a_j(Q)\, dQ = K_{ij}, \qquad \int b_i(Q) f(Q)\, dQ = f_i,$$

then one obtains from the last equation

(4.7) $$C_i = \sum_{j=1}^m K_{ij} C_j + f_i, \qquad i = 1, 2, \cdots, m.$$

Therefore, to every solution of the integral equation (4.2) there corresponds a solution $(C_1, \cdots, C_m)$ of the system (4.7) and the solution is unique because of the linear independence of the functions $a_i(P)$. Conversely, if this system of linear algebraic equations has any solution $(C_1, \cdots, C_m)$, and if we insert this in the right side of (4.5), we obtain a solution of the given integral equation (4.2) because every step leading from (4.2) to (4.7) is reversible. This procedure reduces the problem to an investigation of the system (4.7).

In the same way the transposed equation to (4.2),

(4.8) $$z(P) = \int K(Q, P)\, z(Q)\, dQ + f^*(P),$$

leads to the transposed system to (4.7),

(4.9) $$C_i^* = \sum_{j=1}^m K_{ji} C_j^* + f_i^*, \qquad i = 1, 2, \cdots, m.$$

In virtue of the linear independence of the functions $a_i(P)$ and $b_i(Q)$, to p linearly independent solutions of the homogeneous system (4.7) or (4.9) there correspond exactly p linearly independent solutions of the homogeneous equation (4.2) or (4.8), respectively, and conversely. (Why?) In this manner one achieves a one-to-one correspondence between the solutions of the integral equations (4.2) and (4.8) on the one hand and the solutions of the linear algebraic equations (4.7) and (4.9) on the other. Under this pairing the solutions of the mutually transposed equations (4.2) and (4.8) correspond respectively to the solutions of the mutually transposed equations (4.7) and (4.9).

The first two Fredholm theorems follow directly from this since they hold for the linear algebraic system (4.7). (Verify!)

Let us note the following in order to prove the third Fredholm theorem. If the second case holds for the system (4.7), the condition

$$\sum_{i=1}^{m} f_i C_i^* = 0$$

is necessary and sufficient for the existence of a solution of the system (4.7), where $(C_1^*, \cdots, C_m^*)$ is an arbitrary solution of the transposed homogeneous system. Making use of equation (4.6) this condition may be written in the form

$$\sum_{i=1}^{m} C_i^* \int f(Q) b_i(Q) \, dQ = 0$$

or

(4.10) $$\int f(Q) \left(\sum_{i=1}^{m} C_i^* b_i(Q) \right) dQ = 0.$$

If $(C_1^*, \cdots, C_m^*)$ is a solution of the homogeneous system (4.9), then $\sum C_i^* b_i(Q)$ is a solution of the homogeneous equation (4.8), the transpose of equation (4.2). Hence (4.10) is equivalent to the condition

$$\int f(Q) z(Q) \, dQ = 0$$

for every solution $z(Q)$ of the homogeneous equation (4.8). The third Fredholm theorem for equation (4.2) now follows directly.

REMARKS. 1. It is often the case that the kernel $K(P, Q)$ and the function $f(P)$ are complex functions of the real points P and Q. Then, in general the solutions $y(P)$ of the integral equation (4.2) will also be complex-valued functions of the real point P. All the theorems proved above remain valid in this situation. Let us recall that if

$$\varphi(P) = \varphi_1(P) + i\varphi_2(P),$$

where $\varphi_1(P)$ and $\varphi_2(P)$ are real-valued functions of the real point P, then by definition

$$\int \varphi(P) \, dP = \int \varphi_1(P) \, dP + i \int \varphi_2(P) \, dP.$$

2. Often $a_i(P)$ and $b_i(Q)$ are functions of a complex parameter λ. The

development of this section shows that the first or second Fredholm case holds for equation (4.2) depending on whether the coefficient determinant $D(\lambda)$ of the system (4.7) is zero or not, where the determinant

(4.11) $$D(\lambda) = \begin{vmatrix} 1 - K_{11} & - K_{12} & \cdots & -K_{1m} \\ - K_{21} & 1 - K_{22} & \cdots & -K_{2m} \\ \cdots & \cdots & \cdots & \cdots \\ - K_{m1} & - K_{m2} & \cdots & 1 - K_{mm} \end{vmatrix}$$

with

$$K_{ij} = \int b_i(Q, \lambda) a_j(Q, \lambda) \, dQ.$$

Let the functions $a_j(Q, \lambda)$ and $b_i(Q, \lambda)$ be regular functions of λ in a certain bounded domain Λ of the complex plane for every Q in G. Further it is assumed that $a_j(Q, \lambda)$ and $b_i(Q, \lambda)$ are uniformly continuous functions with respect to both variables Q and λ. Then the K_{ij} and the determinant (4.11) are also regular functions of λ.

That is not hard to show. We represent the integral as a limit of a sum and use the known theorem of Weierstrass that the limit function of a uniformly convergent sequence of regular functions in a domain is also regular in this domain [2]. Hence, the values of λ for which the determinant (4.11) equals zero and consequently those for which the second case holds for (4.2) cannot have a finite limit point in the interior of Λ, provided the determinant (4.11) is different from zero for at least one $\lambda \in \Lambda$.

3. Let K_{ij} and f_i be regular functions of λ for $\lambda \in \Lambda$. This is the case, in particular, if $a_j(Q, \lambda)$ and $b_i(Q, \lambda)$ have the properties enumerated in Remark 2 and $f(Q)$ is a uniformly continuous function of Q, which we shall assume for the sake of simplicity.

According to known theorems from the theory of determinants, the coefficients C_i of the equations (4.7) have the form of fractions with denominator the same determinant (4.11) for all i and with numerators equal to the determinants D_i, which are obtained from (4.11) by replacing the ith column by the column $(f_1, f_2, \cdots, f_m)$. Developing D_i by this last column, we get

$$C_i = \sum_j [M_{ij} f_j / D(\lambda)],$$

where the M_{ij} are polynomials on the K_{ij}. If the above expressions for C_i are inserted in the right side of (4.5), it will follow, making use of (4.6) for the f_i, that:

(4.12) $$y(P) = \int \sum_{ij} [M_{ij} b_j(Q, \lambda) a_i(P, \lambda) f(Q) \, dQ / D(\lambda)] + f(P).$$

The numerator (for each fixed value of P) and the denominator are regular functions of λ in the domain Λ.

It is often useful to write equation (4.12) in the form

(4.13) $$y(P) = \int \bar{\Gamma}(P, Q, \lambda) f(Q) \, dQ + f(P),$$

with

(4.14) $$\bar{\Gamma}(P, Q, \lambda) = \sum_{ij} [M_{ij} b_j(Q, \lambda) a_i(P, \lambda) / D(\lambda)].$$

The function $\bar{\Gamma}(P, Q, \lambda)$ is independent of $f(P)$ and is representable, as shown by (4.14), as a quotient of two regular functions of λ in the domain Λ. $\bar{\Gamma}(P, Q, \lambda)$ can be non-regular in λ only for those values of λ for which $D(\lambda) = 0$. In subsection 2 above it was shown that such values of λ can have no limit point in the interior of Λ provided $D(\lambda)$ is not identically zero, which was expressly assumed. It is easy to show that every such value $\lambda = \lambda_0$, with $D(\lambda_0) = 0$, is really a singular point for $\bar{\Gamma}(P, Q, \lambda)$ in the following sense: $\bar{\Gamma}(P, Q, \lambda)$ is a non-uniformly continuous function of (P, Q, λ) for λ in a sufficiently small neighborhood of λ_0 as P, Q vary over G.

In order to show this we suppose the contrary. Then the function $y(P, \lambda)$ defined by (4.13) will be uniformly continuous for $P \in G$ and λ in a particular neighborhood of λ_0. Now substitute the right side of (4.13), or what is the same thing (4.12), into both sides of (4.2). The results of these insertions will be, for every uniformly continuous function $f(Q)$, uniformly continuous functions of P, λ in the same domains. It is known that these are equal if $\lambda \neq \lambda_0$ and $|\lambda - \lambda_0|$ is sufficiently small, since then $D(\lambda) \neq 0$. Because of continuity these results will agree also for $\lambda = \lambda_0$. Consequently the integral equation (4.2) has a solution at $\lambda = \lambda_0$ for every uniformly continuous function $f(P)$. The solution is given by (4.13) for $\lambda = \lambda_0$, and $\bar{\Gamma}(P, Q, \lambda)$ is defined at $\lambda = \lambda_0$ by continuity. However, for this value of λ the first, but not the second case, applies; consequently $D(\lambda_0) \neq 0$. The foregoing considerations may be easily carried over to the case where $a_i(Q, \lambda)$, $b_i(Q, \lambda)$, and $f(Q)$ have discontinuities with respect to Q, independently of λ, at certain points, sufficiently smooth curves, or surfaces of dimension $\leq (d - 1)$, provided $|a_i(Q, \lambda)|$, $|b_i(Q, \lambda)|$, and $|f(Q)|$ do not grow too rapidly as Q approaches the points of discontinuity. The solution is not determined at those points P where $a_i(P, \lambda)$ and $f(P)$ are not defined.

EXAMPLE.

Let

$$y(x) = -\lambda \int_0^1 (x^2 \xi + x\xi^2) y(\xi) \, d\xi + f(x).$$

Then
$$y(x) = -\lambda \left[x^2 \int_0^1 \xi y(\xi)\, d\xi + x \int_0^1 \xi^2 y(\xi)\, d\xi \right] + f(x).$$

If we set

(4.15) $\qquad \int_0^1 \xi y(\xi)\, d\xi = C_2 \quad \text{and} \quad \int_0^1 \xi^2 y(\xi)\, d\xi = C_1,$

we get

(4.16) $\qquad y(x) = f(x) - C_1 \lambda x - C_2 \lambda x^2.$

If this expression for y is inserted in equation (4.15), it will follow that

$$\int_0^1 \xi [f(\xi) - C_1 \lambda \xi - C_2 \lambda \xi^2]\, d\xi = C_2,$$

$$\int_0^1 \xi^2 [f(\xi) - C_1 \lambda \xi - C_2 \lambda \xi^2]\, d\xi = C_1$$

or

(4.17) $\quad b_1 - C_1 \lambda/3 - C_2 \lambda/4 = C_2, \qquad b_2 - C_1 \lambda/4 - C_2 \lambda/5 = C_1,$

where

$$b_1 = \int_0^1 \xi f(\xi)\, d\xi \quad \text{and} \quad b_2 = \int_0^1 \xi^2 f(\xi)\, d\xi.$$

We now write equations (4.17) in the form

(4.18) $\qquad \begin{aligned} C_1 \lambda/3 + C_2(1 + \lambda/4) &= b_1, \\ C_1(1 + \lambda/4) + C_2 \lambda/5 &= b_2. \end{aligned}$

The determinant of this system is equal to

$$1 + \lambda/2 - \lambda^2/240$$

with the two roots

$$\lambda = 60 \pm 16(15^{\frac{1}{2}}).$$

Only for these two values of λ does the second Fredholm case enter. Here all the solutions of the homogeneous integral equation

$$y(x) + \lambda \int_0^1 (x^2 \xi + x \xi^2) y(\xi)\, d\xi = 0$$

are given by

$$y(x) = C[x \mp 5(15^{-\frac{1}{2}}) x^2],$$

where C is an arbitrary constant. For the other values of λ the integral equation has a unique solution given by formula (4.16), where C_1 and C_2 are uniquely determined by (4.18). This solution may be written in the form (4.13), where

$$\bar{\Gamma}(x, \xi, \lambda) = \lambda[(\xi x/5)\lambda - (1 + \lambda/4)\xi^2 x$$
$$- \xi x^2(1 + \lambda/4) + \xi^2 x^2 \lambda/3]/(1 + \lambda/2 - \lambda^2/240).$$

EXERCISES. Determine $u(x)$ from the following equations:

1. $u(x) = e^x + \lambda \int_0^{10} xtu(t)\,dt.$

2. $u(x) = \lambda \int_0^{\pi} \sin x\, u(t)\,dt.$

3. $u(x) = \lambda \int_0^{\pi} \cos x\, u(t)\,dt.$

4. $u(x) = x + \lambda \int_0^1 (x - t)u(t)\,dt.$

5. $u(x) = \lambda \int_0^{2\pi} \sin x \sin t\, u(t)\,dt + f(x).$

§5. Integral equations with continuous kernels of sufficiently small absolute bound

For such equations the first case always holds, i.e. they always have a unique solution. This can be shown by the method of successive approximation. As in the theory of ordinary differential equations, one demonstrates the existence and uniqueness of the solution of an integral equation that is equivalent to a given differential equation with initial conditions. This is essentially an application of the principle of contraction mappings [3]. Here I can only show that it is possible to apply this general principle. For this reason I show how to carry out the proof for a given concrete case since one obtains thereby specific formulae that will be useful in the sequel.

First we introduce a symbolic notation that we shall occasionally use in the sequel. Let $K_1(P, Q)$ and $K_2(P, Q)$ be uniformly continuous functions of P and Q for $P \in G$ and $Q \in G$. Let us set

(5.1) $$K_2 \circ K_1 = \int K_2(P, S) K_1(S, Q)\,dS.$$

We call the kernel $K(P, Q) = K_2 \circ K_1$ the *symbolic product* [usually called the *convolution*-Trans.] of the kernels $K_2(P, Q)$ and $K_1(P, Q)$.

The symbolic multiplication of kernels introduced is analogous to the multiplication of two matrices.

The function $\varphi_1(P)$ is transformed into the function

$$\varphi_2(P) = \int K_1(P, Q)\varphi_1(Q) \, dQ$$

by means of the kernel $K_1(P, Q)$, and the function $\varphi_2(P)$ into the function $\varphi_3(P) = \int K_2(P, Q)\varphi_2(Q) \, dQ$ by the kernel $K_2(P, Q)$. Then the kernel $K_2 \circ K_1$ yields the transformation of the function $\varphi_1(P)$ into the function $\varphi_3(P)$, i.e. $\varphi_3(P) = \int (K_2 \circ K_1)\varphi_1(Q) \, dQ$. In exactly the same fashion, in m-dimensional space, the successive application of two linear transformations yields a linear transformation with a matrix that is the product of the matrices of these linear transformations. One easily shows as follows that $K_2 \circ K_1$ itself is again a uniformly continuous function of P and Q.

(5.2)
$$\left| \int K_2(P_1, S)K_1(S, Q_1) \, dS - \int K_2(P_2, S)K_1(S, Q_2) \, dS \right|$$
$$\leq \left| \int K_2(P_1, S)[K_1(S, Q_1) - K_1(S, Q_2)] \, dS \right|$$
$$+ \left| \int K_1(S, Q_2)[K_2(P_1, S) - K_2(P_2, S)] \, dS \right|.$$

Suppose the upper bounds of the absolute values of $K_1(P, Q)$ and $K_2(P, Q)$ do not exceed M for $P \in G$ and $Q \in G$. Let D be the volume of the domain G. Because of the uniform continuity of $K_1(P, Q)$ and $K_2(P, Q)$, for every $\epsilon > 0$ there is an $\eta > 0$ such that

$$| K_2(P_1, S) - K_2(P_2, S) | < \epsilon/2DM$$

and

$$| K_1(S, Q_1) - K_1(S, Q_2) | < \epsilon/2DM,$$

provided the distance between the points P_1 and P_2 and between the points Q_1 and Q_2 is less than η. One now easily sees that under these conditions the left half of the inequality (5.2) is less than ϵ. Let us note that in general $K_2 \circ K_1 \neq K_1 \circ K_2$. If $K_3(P, Q)$ is also a uniformly continuous function of P and Q, it is easy to prove that

$$K_1 \circ (K_2 \circ K_3) = (K_1 \circ K_2) \circ K_3.$$

After these preliminary remarks we shall show that an integral equation with a continuous kernel of sufficiently small absolute value always possesses a unique solution. This will be used later to demonstrate the Fredholm theorems for integral equations with arbitrary continuous kernels.

Let the given integral equation be

§5] CONTINUOUS KERNELS OF SMALL BOUND

(5.3) $$y(P) = \lambda \int K(P, Q) y(Q) \, dQ + f(P),$$

where $K(P, Q)$ and $f(P)$ are uniformly continuous functions for $P \in G$ and $Q \in G$ and G is a bounded domain. Here λ is a parameter. It is ordinarily contained in the equation in the form of (5.3).

Instead of emphasizing each time the uniform continuity of the functions in an open region G, one may consider these functions in a bounded closed region $\bar{G}$ (i.e. G together with its boundary) and only require their continuity. Then the uniform continuity of these functions will follow. If a uniformly continuous function φ is given in an open region G, one may extend it to the boundary of the region because of continuity. One then has a uniformly continuous function in the closed region $\bar{G}$. Those simple regions that we shall consider (cf. the Remark to §1) have boundaries whose d-dimensional volume is zero. Thus, the integral of the function over the region G agrees with the integral of its extension over $\bar{G}$.

All further considerations of this section hold equally well whether the functions considered take on real or complex values. Also the parameter λ may be complex. It is essential however that the points P and Q be real, i.e. that all the coordinates of these points be real. Otherwise it becomes necessary to define what is meant by an integral of several complex variables.

It follows from the definition of the kernel given before that one should now call $\lambda K(P, Q)$ the kernel. Following the general terminology we also call the function $K(P, Q)$ the kernel of the integral equation (5.3). But in the heading of this section the statement about the smallness of the kernel refers to the value of $\lambda K(P, Q)$.

We now seek a solution of the integral equation (5.3) in the form of an infinite power series in λ:

(5.4) $$y(P) = y_0(P) + \lambda y_1(P) + \lambda^2 y_2(P) + \cdots.$$

Upon inserting this series formally in (5.3) it follows that:

(5.5) $$\begin{aligned} y_0(P) + \lambda y_1(P) + \cdots \\ = \lambda \int K(P, Q)[y_0(Q) + \lambda y_1(Q) + \cdots] \, dQ + f(P). \end{aligned}$$

If we equate the coefficients of like powers of λ, we get

(5.6) $$\begin{aligned} y_0(P) &= f(P), \\ y_{k+1}(P) &= \int K(P, Q) y_k(Q) \, dQ, \quad k = 0, 1, 2, \cdots, \end{aligned}$$

or

$$y_0(P) = f(P),$$

$$y_1(P) = \int K(P, P_1)f(P_1)\, dP_1,$$

$$y_2(P) = \iint K(P, P_1)K(P_1, P_2)f(P_2)\, dP_1\, dP_2,$$

$$\cdots\cdots\cdots\cdots\cdots\cdots\cdots\cdots\cdots\cdots\cdots\cdots\cdots$$

$$y_k(P) =$$

(5.7)
$$(k\text{-fold}) \int \cdots \int K(P, P_1) \cdots K(P_{k-1}, P_k)f(P_k)\, dP_1 \cdots dP_k.$$

This last equation may also be written in the following form:

(5.8) $$y_k(P) = \int K^{(k)}(P, Q)f(Q)\, dQ, \qquad k = 1, 2, 3, \cdots,$$

where

$$K^{(k)}(P, Q) =$$

(5.9) $$((k-1)\text{-fold}) \int \cdots \int K(P, P_1) \cdots K(P_{k-1}, Q)\, dP_1 \cdots dP_{k-1},$$

$$k = 2, 3, \cdots,$$

and

$$K^{(1)}(P, Q) = K(P, Q).$$

Using the symbolic notation we can also represent the kernel $K^{(k)}(P, Q)$ in the form

(5.10) $$K^{(k)}(P, Q) = K \circ K \circ \cdots \circ K\, (k\text{-times}).$$

As we showed at the beginning of this section all the kernels $K^{(k)}(P, Q)$ are uniformly continuous. The function $K^{(k)}(P, Q)$ is called the kth iterated kernel of $K(P, Q)$. One easily sees that all the $y_k(P)$ are likewise uniformly continuous.

We now estimate the kernel $K^{(k)}(P, Q)$. Because of uniform continuity the kernel $K(P, Q)$ is bounded. Suppose

(5.11) $$|K(P, Q)| < M.$$

From this estimate it follows from (5.9) that

(5.12) $$|K^{(k)}(P, Q)| \leq M^k D^{k-1},$$

where D is the volume of the region G. From this it follows because of (5.8) that

$$|y_k(P)| \leq M^k D^k F,$$

where F is the upper bound of $|f(P)|$. Therefore the series (5.4) will converge absolutely and uniformly for $P \in G$ if

(5.13) $$|\lambda| < 1/MD.$$

Since every term of the series is continuous, the sum of the series will also be a continuous function of P. Because the series (5.4) converges uniformly we can carry out term by term the integration indicated in the above formally written equation (5.5). The definition of $y_k(P)$ by means of the formulae (5.6) implies that equation (5.5) really holds, i.e. that the function $y(P)$ defined by the series (5.4) is a solution of the integral equation (5.3).

We now show that this solution is the only one in the class of bounded functions for which the condition (5.13) obtains. Let us suppose that there are two solutions y_1, y_2 of (5.3). Substituting these into (5.3), we get upon subtraction the equation

(5.14) $$y_2(P) - y_1(P) = \lambda \int K(P, Q)[y_2(Q) - y_1(Q)] \, dQ.$$

Designating the upper bound of $|y_2(P) - y_1(P)|$ by Y, one obtains from (5.14), making use of (5.11), the inequality

$$Y \leq |\lambda| MDY.$$

From (5.13) it follows that

$$Y \leq cY, \quad \text{with } c < 1.$$

This is possible only if $Y = 0$, q.e.d.

It is often convenient to represent the solution of the integral equation (5.3) in the form

(5.15) $$y(P) = \lambda \int \Gamma(P, Q, \lambda) f(Q) \, dQ + f(P),$$

with

(5.16) $$\Gamma(P, Q, \lambda) = \sum_{k=1}^{\infty} \lambda^{k-1} K^{(k)}(P, Q).$$

From the estimate (5.12) it follows that the series (5.16) converges uniformly with respect to (P, Q, λ) if $P \in G$, $Q \in G$ and $|\lambda| < (1/MD) - \epsilon$, $\epsilon > 0$. Thus $\Gamma(P, Q, \lambda)$ is a uniformly continuous function of P, Q for fixed λ, and a regular function of λ in the circle (5.13) for $P \in G$, $Q \subset G$. Hence the integral (5.15) exists. It is easy to see that it indeed yields the solution of the integral equation (5.3) represented by the series (5.4), if one substitutes the series (5.16) in place of $\Gamma(P, Q, \lambda)$ in the right hand side of (5.15) and integrates termwise with respect to Q.

The function $\Gamma(P, Q, \lambda)$ is called the *resolvent* of the integral equation (5.3).

We compare (5.15) with (4.13). We shall now show that for the integral equation (5.3) with a degenerate kernel, for which $a_i(P)$ and $b_i(P)$ are uniformly continuous and have sufficiently small absolute values, i.e. for any integral equation that belongs simultaneously to the types considered in §§4 and 5, the equation

$$\bar\Gamma(P, Q, \lambda) = \lambda \Gamma(P, Q, \lambda)$$

holds. Since we are dealing here with the first case, $D(\lambda) \neq 0$.

We assume that

$$\bar\Gamma(P_0, Q_0, \lambda_0) \neq \lambda_0 \Gamma(P_0, Q_0, \lambda_0), \qquad D(\lambda_0) \neq 0,$$

at a point (P_0, Q_0, λ_0). Since the kernels $\bar\Gamma(P_0, Q, \lambda_0)$ and $\Gamma(P_0, Q, \lambda_0)$ are continuous in Q, there exists a neighborhood G_0 of the point Q_0 in which

$$\mathrm{Re}\,\{\bar\Gamma(P_0, Q, \lambda_0)\} \neq \mathrm{Re}\,\{\lambda_0 \Gamma(P_0, Q, \lambda_0)\}$$

or

$$\mathrm{Im}\,\{\bar\Gamma(P_0, Q, \lambda_0)\} \neq \mathrm{Im}\,\{\lambda_0 \Gamma(P_0, Q, \lambda_0)\}$$

everywhere. On the other hand

$$\int \bar\Gamma(P_0, Q, \lambda_0) f(Q)\, dQ = \lambda_0 \int \Gamma(P_0, Q, \lambda_0) f(Q)\, dQ$$

must hold for an arbitrary uniformly continuous function $f(P)$ because of the uniqueness of the solution of integral equations of the type considered. In particular, this equation must hold for a function $f(Q)$ which is positive everywhere in the interior of the neighborhood G_0 of the point Q_0 and zero outside this neighborhood. This is not possible. (Why?)

As is evident from (5.16), $\Gamma(P, Q, \lambda)$ is determined by the kernel of the integral equation and is independent of $f(P)$. Since the function $y(P)$ defined by (5.15) represents the only solution of (5.3), it follows that the equations (5.3) and (5.15) are equivalent. If one therefore takes the function $y(P)$ in the equation (5.15) as known and $f(P)$ as unknown, then the only solution $f(P)$ of this equation will be furnished by the formula (5.3). The function $K(P, Q)$ plays in this formula the role of the resolvent for the equation (5.15) with the kernel $\Gamma(P, Q, \lambda)$.

If the same considerations as above are applied to the transposed equation

(5.17) $$z(P) = \lambda \int K(Q, P)z(Q)\, dQ + f(P)$$

of equation (5.3), it follows that (5.17) has a unique solution in the circle (5.13) in the class of bounded functions, which will be furnished by the series

$$z(P) = z_0(P) + \lambda z_1(P) + \lambda^2 z_2(P) + \cdots .$$

Here

$$z_0(P) = f(P),$$

$$z_k(P) = \int K(Q, P)z_{k-1}(Q)\, dQ.$$

If the kernel $K(Q, P)$ is symbolized by $K^*(P, Q)$, it will follow that

$$z_1(P) = \int K^*(P, P_1)f(P_1)\, dP_1 ,$$

$z_k(P) =$

(k-fold) $\int \cdots \int K^*(P, P_1)K^*(P_1, P_2) \cdots K^*(P_{k-1}, P_k)f(P_k)\, dP_1 \cdots dP_k$

or

$$z_k(P) = \int K^{*(k)}(P, Q)f(Q)\, dQ, \qquad k = 1, 2, \cdots ,$$

with

$K^{*(k)}(P, Q) =$

$((k-1)$-fold$) \int \cdots \int K^*(P, P_1) \cdots K^*(P_{k-1}, Q)\, dP_1 \cdots dP_{k-1} .$

If one writes out the corresponding integrals, one sees that

$$K^{*(k)}(P, Q) = K^{(k)}(Q, P).$$

From this it follows that the solution of (5.17) has the form

(5.18) $$z(P) = \lambda \int \Gamma^*(P, Q, \lambda)f(Q)\, dQ + f(P),$$

with

$$\Gamma^*(P, Q, \lambda) - \Gamma(Q, P, \lambda).$$

We have thus shown, in this way, that equation (5.3) and also its transposed equation (5.17) with the agreed upon assumptions on the kernel have a unique solution in the circle (5.13) for every uniformly continuous function $f(P)$, i.e. it has been demonstrated that here the first Fredholm case holds.

In conclusion we point out the following two formulae:

$$(5.19) \quad \Gamma(P, Q, \lambda) = K(P, Q) + \lambda \int K(P, P_1) \Gamma(P_1, Q, \lambda) \, dP_1,$$

$$(5.20) \quad \Gamma(P, Q, \lambda) = K(P, Q) + \lambda \int \Gamma(P, P_1, \lambda) K(P_1, Q) \, dP_1.$$

The validity of these formulae is shown immediately if one substitutes for Γ the series (5.16) and compares the coefficients of like powers of λ, making use of the formula (5.10).

For sufficiently small $|\lambda|$ one may use the method of successive approximations for the approximate computation of the solution of the integral equation [4].

§6. Integral equations with almost degenerate kernels

Consider the integral equation

$$(6.1) \quad y(P) = \lambda \int K(P, Q) y(Q) \, dQ + f(P),$$

where $f(P)$ is a uniformly continuous function and

$$K(P, Q) = \sum_{i=1}^{m} a_i(P) b_i(Q) + K_1(P, Q) = A(P, Q) + K_1(P, Q).$$

Here $a_i(P)$, $b_i(Q)$ and $K_1(P, Q)$ are uniformly continuous and, since their domain of definition is bounded, they are also bounded functions. It is again immaterial whether these functions take on complex or only real values. In order not to obscure the essence of the matter, we shall write the integral equation in symbolic form. The equation

$$(6.2) \quad \psi(P) = \int K(P, Q) y(Q) \, dQ$$

will be written symbolically in the form

$$\psi = Ky.$$

K is thus an operator that transforms the function $y(P)$ into the function $\psi(P) = \int K(P, Q) y(Q) \, dQ$. This operator is defined by the kernel $K(P, Q)$. We designate by K^* the operator defined by the transposed kernel $K^*(P, Q) = K(Q, P)$. Finally let E be the operator that transforms every function $y(P)$ into itself, i.e. $Ey = y$ for every function $y(P)$. The operator $K_1 \pm K_2$ is defined by the equation $(K_1 \pm K_2)y = K_1y \pm K_2y$. The operator K_1K_2 is defined by $(K_1K_2)y = K_1(K_2y)$ for an arbitrary function $y(P)$.

It is easy to see that the operator $K_1 \pm K_2$ is defined by the kernel $K_1(P, Q) \pm K_2(P, Q)$ and the operator K_1K_2 by the kernel $K_1 \circ K_2$ if K_1 and K_2 are operators of the form (6.2) with the kernels $K_1(P, Q)$ and $K_2(P, Q)$.

Accordingly one may now write the equation (6.1) in the form

$$(E - \lambda K)y = f.$$

Before we turn to the proof of the Fredholm theorems for equation (6.1) we first formulate the following lemmas:

LEMMA 1. *If $A(P, Q)$ is a degenerate kernel and $K(P, Q)$ an arbitrary continuous kernel, then $A \circ K$ and $K \circ A$ are also degenerate kernels.*

LEMMA 2. *The transpose of the kernel $K_1 \circ K_2$ is equal to $K_2^* \circ K_1^*$.*

The correctness of these assertions follows immediately from the consideration of the corresponding integrals.

Now we shall prove that the three Fredholm theorems hold for equation (6.1) for $|\lambda| < 1/M_1D$, where M_1 is the upper bound for $|K_1(P, Q)|$ and D is the volume of G.

1. *The first Fredholm theorem.* We shall show that the non-homogeneous equation (6.1) has a solution for every function $f(P)$, provided the corresponding homogeneous equation to (6.1) has only a trivial solution.

Replacing K by $A + K_1$ we shall rewrite equation (6.1) in the form

$$(E - \lambda A - \lambda K_1)y = f,$$

where A and K_1 are operators with the kernels $A(P, Q)$ and $K_1(P, Q)$ respectively. Then

(6.3) $$(E - \lambda K_1)y = \lambda Ay + f.$$

Let us put

(6.4) $$(E - \lambda K_1)y = \eta.$$

Since $|\lambda| < 1/M_1D$, it follows from the formula (5.15) demonstrated in the previous section that

(6.5) $$y = \eta + \lambda \Gamma \eta = (E + \lambda \Gamma)\eta,$$

where Γ is an operator corresponding to the resolvent $\Gamma(P, Q, \lambda)$ of the kernel $K_1(P, Q)$. If this expression for $y(P)$ is inserted in equation (6.3), one obtains

$$\eta = \lambda A(E + \lambda \Gamma)\eta + f$$

or

(6.6) $$[E - \lambda A(E + \lambda \Gamma)]\eta = f.$$

The kernel $A(P, Q) + A \circ \lambda\Gamma(P, Q)$ of this integral equation is degenerate by Lemma 1. In this way we have shown that to every solution $y(P)$ of equation (6.1) there corresponds by (6.4) a solution $\eta(P)$ of equation (6.6) with degenerate kernel.

Conversely one easily proves that to every solution $\eta(P)$ of equation (6.6) there corresponds a solution $y(P)$ of equation (6.1) defined by formula (6.5).

Since by assumption the homogeneous equation (6.1) has only a trivial solution, it follows that the homogeneous equation (6.6) also has only a trivial solution.

But for equation (6.6) with degenerate kernel, the first Fredholm theorem was demonstrated in §4. Hence the non-homogeneous equation (6.6) possesses a solution $\eta(P)$ for every function $f(P)$. And from formula (6.5) one obtains a solution $y(P)$ for equation (6.1) for an arbitrary function $f(P)$. Clearly this is the only solution.

We have thus demonstrated the first Fredholm theorem, since the non-homogeneous equation either has none or more than one solution whenever the homogeneous equation has a non-trivial solution.

2. *The second Fredholm theorem.* We shall show that the equation $(E - \lambda A - \lambda K_1)y = 0$ and its transposed equation

(6.7) $$(E - \lambda A^* - \lambda K_1^*)z = 0$$

have the same number of linearly independent solutions if $|\lambda| < 1/M_1 D$.

First we remark that the homogeneous equations (6.1) and (6.6) have the same number of linearly independent solutions, since to every p linearly independent solutions of one of these equations there correspond also p linearly independent solutions of the other, by (6.4) or (6.5).

The transposed homogeneous equation to (6.6) has the form

(6.8) $$[E - \lambda(E + \lambda\Gamma^*)A^*]\zeta = 0$$

by Lemma 2. Since equation (6.6) has a degenerate kernel, the homogeneous equations (6.6) and (6.8) have the same number of linearly independent solutions by the second Fredholm theorem demonstrated in §4 for equations with degenerate kernels. Let us show that equations (6.8) and (6.7) are equivalent. Suppose that $\zeta(P)$ is a solution of (6.8). We shall prove that it also satisfies (6.7). If we apply the operator $E - \lambda K_1^*$ to both sides of (6.8), we obtain

(6.9)
$$(E - \lambda K_1^*)[E - \lambda(E + \lambda\Gamma^*)A^*]\zeta$$
$$= [E - \lambda K_1^* - \lambda(E - \lambda K_1^*)(E + \lambda\Gamma^*)A^*]\zeta = 0.$$

Since it follows from formulae (5.17) and (5.18) that for an arbitrary function $\varphi(P)$

$$(E - \lambda K_1^*)(E + \lambda\Gamma^*)\varphi = \varphi,$$

we get from equation (6.9)

$$(E - \lambda K_1^* - \lambda A^*)\zeta = 0.$$

If one applies, analogously, the operator $E + \lambda\Gamma^*$ to equation (6.7) and makes use of the equation $(E + \lambda\Gamma^*)(E - \lambda K_1^*) = E$, it will follow that every solution of (6.7) also satisfies (6.8). Hence, we have shown that the homogeneous equations (6.1), (6.6), (6.8), and (6.7) have the same number of linearly independent solutions. This proves the second Fredholm theorem.

Those λ values for which the second Fredholm case applies for equation (6.1) we call the *eigenvalues* of equation (6.1) [or of the kernel $K(P, Q)$, cf. §2] and the corresponding non-trivial solutions of the homogeneous equation for each such λ the *eigenfunctions belonging to this eigenvalue*.

Since the function $\Gamma(P, Q, \lambda)$ is a regular function of λ in the circle $|\lambda| < 1/M_1D$, the corresponding determinant (4.11) for the degenerate equaton (6.6) is also a regular function of λ in this circle. Since this determinant is 1 for $\lambda = 0$, it cannot be identically zero. Consequently its roots cannot have a limit point in this circle. So we have shown that *the eigenvalues λ of equation* (6.1) *cannot have a limit point in the circle* $|\lambda| < 1/M_1D$.

3. *The third Fredholm theorem.* We shall now show that equation (6.1) has a solution if, and only if,

$$\int f(P)z(P)\,dP = 0,$$

where $z(P)$ is any solution of the transposed homogeneous equation (6.7) of equation (6.1).

In the proof of the first Fredholm theorem for equation (6.1) we established that equation (6.1) has a solution if, and only if, equation (6.6) with degenerate kernel has one. In §4 we showed that equation (6.6) with degenerate kernel has a solution if, and only if,

$$\int f(P)\zeta(P)\,dP = 0,$$

where $\zeta(P)$ is an arbitrary solution of equation (6.8). But according to what was shown above, the totality of these solutions $\zeta(P)$ coincides with the set of all solutions $z(P)$ of equation (6.7), demonstrating the theorem.

§7. Integral equations with uniformly continuous kernels

Every uniformly continuous kernel $K(P, Q)$ can be uniformly approximated by a degenerate kernel with arbitrary accuracy. Let $K(P, Q)$ be a uniformly continuous function of (P, Q) defined on a bounded domain G. According to the Weierstrass theorem [5], there exists for every $\epsilon > 0$ a polynomial $K_0(P, Q)$ of sufficiently high order in P and Q such that the inequality

$$| K(P, Q) - K_0(P, Q) | < \epsilon$$

holds everywhere in G. It is clear that one can represent every term of the polynomial $K_0(P, Q)$ in the form of a product of two factors, one of which depends only on P, the other only on Q. Consequently one may write

$$K(P, Q) = \sum_{i=1}^{N} a_i(P) b_i(Q) + K_1(P, Q),$$

with

$$| K_1(P, Q) | < \epsilon.$$

Using the theorem proved in the previous section it follows that all three Fredholm theorems hold in the circle

$$|\lambda| < 1/\epsilon D;$$

D is again the volume of G, and the eigenvalues have no limit point in this circle. Since ϵ may be chosen arbitrarily small, the correctness of the three theorems follows in arbitrarily large circles about the midpoint $\lambda = 0$, i.e. in the whole λ-plane.

We recapitulate the course of the considerations that have led to the proof of the Fredholm theorems for equations with uniformly continuous kernels. First (§4) these theorems were shown for integral equations with degenerate kernels. In §6 we extended these theorems to integral equations with almost degenerate kernels. In this section we showed that every uniformly continuous kernel may be uniformly approximated with arbitrary accuracy by a degenerate kernel. From this we gained a proof of the Fredholm theorems for integral equations with arbitrary uniformly continuous kernels.

This method of proof for the Fredholm theorems originated with E. Schmidt (in the exposition of these theorems I have made use of the lecture notes of S. L. Sobolev). It should be noted that one may obtain approximate solutions of integral equations with continuous kernels by approximating these kernels with degenerate ones [6].

§8. Integral equations with kernels of the form $\bar{K}(P, Q)/PQ^\alpha$

1. Now let the points P, Q belong to a specific closed and bounded domain $\bar{G}$ (cf. remarks on p. 19) and let $\bar{K}(P, Q)$ be a continuous function of (P, Q). PQ is the distance between P and Q. The purpose of subsection 1 of this section is to prove that *for integral equations with kernels of this kind for $\alpha < d$, where d is the dimension of the domain G, the three Fredholm theorems hold in the whole λ-plane and that in this case the eigenvalues cannot have a finite limit point.*

As preparation we shall show first the following lemma for the kernels $K_1(P, Q)$ and $K_2(P, Q)$, continuous with respect to (P, Q), if $P \in G$ and $Q \in G$ and $P \neq Q$.

If

(8.1) $\qquad |K_1(P, Q)| < A_1/PQ^\alpha, \qquad 0 \le \alpha < d,$

and

(8.2) $\qquad |K_2(P, Q)| < A_2/PQ^\beta, \qquad 0 \le \beta < d,$

then the integral

$$K_3(P, Q) = \int K_1(P, P_1) K_2(P_1, Q) \, dP_1$$

always exists and is continuous with respect to (P, Q) when P is distinct from Q. Further,

(8.3) $\qquad |K_3(P, Q)| < A_3/PQ^{\alpha+\beta-d}, \qquad \text{if } \alpha + \beta > d,$

and

(8.4) $\qquad |K_3(P, Q)| < A_3 |\ln PQ| + A_4, \qquad \text{if } \alpha + \beta = d,$

where A_3 and A_4 are constants. If however $\alpha + \beta < d$, then this integral always exists and is a uniformly continuous function of (P, Q).

Proof. If $P \neq Q$, then

$$|K_3(P, Q)| \le \int |K_1(P, P_1)| \, |K_2(P_1, Q)| \, dP_1$$

(8.5) $\quad \le (d\text{-fold}) \displaystyle\int \cdots \int_{r \le D} \dfrac{A_1 A_2 \, dx_1^{(1)} \cdots dx_d^{(1)}}{\left[\sum_{i=1}^{d}(x_i - x_i^{(1)})^2\right]^{\alpha/2} \left[\sum_{i=1}^{d}(x_i^{(1)} - y_i)^2\right]^{\beta/2}}.$

In this equation x_i, $x_i^{(1)}$, y_i, $i = 1, \cdots, d$, stand for the corresponding coordinates of the points P, P_1, Q; D the diameter of the domain $\bar{G}$, i.e.

the least upper bound of the distances between two of its points; and $r = [\sum (x_i^{(1)} - x_i)^2]^{\frac{1}{2}}$.

Without loss of generality we put $x_1 = \cdots = x_d = 0$, $y_1 = \rho$, $y_2 = \cdots = y_d = 0$, $PQ = \rho$ to simplify the computation, and further let

$$x_i^{(1)} = \rho \xi_i.$$

Then one may write the integral I on the right side of (8.5) as

$$I = (d\text{-fold}) \int \cdots \int_{r \leq D} \frac{A_1 A_2 \rho^d \, d\xi_1 \cdots d\xi_d}{\left(\sum_{i=1}^d \xi_i^2\right)^{\alpha/2} \left[(\xi_1 - 1)^2 + \sum_{i=2}^d \xi_i^2\right]^{\beta/2}} \rho^{\alpha + \beta}.$$

Let us note that if $(\sum_{i=1}^d \xi_i^2)^{\frac{1}{2}} \geq 2$, then

(8.6) $$\left[(\xi_1 - 1)^2 + \sum_{i=2}^d \xi_i^2\right]^{\frac{1}{2}} \geq \frac{1}{2} \left[\sum_{i=1}^d \xi_i^2\right]^{\frac{1}{2}}.$$

From Fig. 2 it is clear that $PM + OP \geq OM$. But $OP = 1$. Consequently $PM \geq OM - 1 = \frac{1}{2}(OM + (OM - 2))$. Since by assumption $OM \geq 2$, $PM \geq OM/2$.

Now the integral I is split in two parts and the estimate (8.6) used.

$$I \leq \frac{A_1 A_2}{\rho^{\alpha + \beta - d}} \int \cdots \int_{\Sigma \xi_i^2 \leq 4} \frac{d\xi_1 \cdots d\xi_d}{\left[\sum_{i=1}^d \xi_i^2\right]^{\alpha/2} \left[(\xi_1 - 1)^2 + \sum_{i=2}^d \xi_i^2\right]^{\beta/2}}$$

$$+ \frac{A_1 A_2}{\rho^{\alpha + \beta - d}} \int \cdots \int_{4 \leq \Sigma \xi_i^2 \leq D^2/\rho^2} \frac{2^\beta \, d\xi_1 \cdots d\xi_d}{\left[\sum_{i=1}^d \xi_i^2\right]^{(\alpha + \beta)/2}}.$$

The integral over the first range is convergent to a constant C_1 independent of ρ. To compute the second integral we go over to polar coordinates:

(8.7) $$I \leq C_1 \rho^{d - \alpha - \beta} + C_2 \rho^{d - \alpha - \beta} \int_2^{D/\rho} \tau^{d - 1 - \alpha - \beta} \, d\tau,$$

where C_2 is a positive constant.

For $\alpha + \beta > d$ it follows from the last formula that

$$I \leq C_1 \rho^{d - \alpha - \beta} + C_2 \rho^{d - \alpha - \beta} \int_2^\infty \tau^{d - 1 - \alpha - \beta} \, d\tau = C_3 \rho^{d - \alpha - \beta};$$

thus the estimate (8.3).

It follows from (8.7) for $\alpha + \beta = d$ that

$$I \leq C_1 + C_2 \ln (D/2\rho),$$

which is the estimate (8.4) for $K_3(P, Q)$. In the sequel, the case $\alpha + \beta = d$ will be avoided by increasing α and β.

If however $\alpha + \beta < d$, then it is clear, that $K_3(P, Q)$ exists for $P = Q$. And further it follows from the estimate (8.7) that:

(8.8) $\quad I \leq C_1 \rho^{d-\alpha-\beta} + C_2(\rho^{d-\alpha-\beta}/d - \alpha - \beta) \, [(D/\rho)^{d-\alpha-\beta} - 2^{d-\alpha-\beta}] \leq C_3$

for some constant C_3.

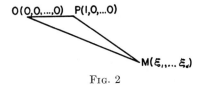

Fig. 2

We shall now show that $K_3(P, Q)$ always depends continuously on (P, Q), if P does not coincide with Q. To this end we note that

$$|K_3(P, Q) - K_3(P^*, Q^*)|$$

$$\leq |K_3(P, Q) - K_3(P, Q^*)| + |K_3(P, Q^*) - K_3(P^*, Q^*)|$$

(8.9)

$$\leq \int |K_1(P, P_1)| \, |K_2(P_1, Q) - K_2(P_1, Q^*)| \, dP_1$$

$$+ \int |K_2(P_1, Q^*)| \, |K_1(P, P_1) - K_1(P^*, P_1)| \, dP.$$

It is now assumed that the functions $K_1(P, Q)$ and $K_2(P, Q)$ are defined for all points P and Q belonging to $\bar{G}(P \neq Q)$ and are continuous everywhere where $P \neq Q$. Hence the functions $K_1(P, Q)$ and $K_2(P, Q)$ are uniformly continuous with respect to (P, Q) in an arbitrary closed set of points (P, Q) that contains no points for which $P = Q$. Thus, each difference under the integral signs is uniformly small with respect to P_1 if only the points Q and Q^*, P and P^* lie sufficiently close to one another for the whole domain $\bar{G}$ of P_1 with the exception of certain neighborhoods G_1, G_2, G_3, G_4 of the points $P_1 = Q$, $P_1 = Q^*$, $P_1 = P$, $P_1 = P^*$. We include in G_1, G_2, G_3, G_4 the points of G which lie no further from Q, Q^*, P, P^* than a small fixed distance r which does not change as P approaches P^* or Q approaches Q^*. Because of conditions (8.1) and (8.2), the integrals in (8.9) extended over $\bar{G} - (G_1 + G_2)$ and $\bar{G} - (G_3 + G_4)$ will become arbitrarily small for sufficiently near approach of the points (P, P^*) and (Q, Q^*). The part of the integrals (8.9) which are taken over the neighborhoods G_1, G_2, G_3, G_4 will become arbitrarily small for $P \neq Q$ as $r \to 0$ because of conditions (8.1) and (8.2).

If however $\alpha + \beta < d$, the integrals (8.9) will become arbitrarily small for sufficient proximity of the points (P, Q) and (P^*, Q^*) even if the points P and Q (or P^* and Q^*) coincide, since in this case the part of the integrals over the neighborhoods G_1, G_2, G_3, G_4 tend to zero as $r \to 0$ uniformly with respect to (P, Q).

Indeed, the first of the integrals (8.9) does not exceed for these neighborhoods the sum

$$\int |K_1(P, P_1)| \, |K_2(P_1, Q)| \, dP_1 + \int |K_1(P, P_1)| \, |K_2(P_1, Q^*)| \, dP_1.$$

Each of these integrals is estimated by taking into consideration the inequality (8.8). One deals analogously with the second of the integrals (8.9) extended over G_1, G_2, G_3, G_4.

The continuity of the function $K_3(P, Q)$ follows in its entire closed domain, and hence also its uniform continuity.

We consider now integral equations

$$(8.10) \qquad y(P) = \lambda \int K(P, Q) y(Q) \, dQ + f(P),$$

where $K(P, Q)$ has the form indicated in the section heading and $\alpha < d$. The function $f(P)$ is to be continuous in the closed domain and therefore also bounded. Further, we shall consider only continuous solutions. We remark only that one may easily show exactly as above that under the assumptions on $K(P, Q)$ every bounded solution of equation (8.10) is continuous if $f(P)$ is continuous.

We shall first show that this equation as well as its transpose always has a unique solution in the class of bounded functions for sufficiently small $|\lambda|$. Since, all proofs for the transposed equation follow exactly as for the given equation, we confine ourselves to equation (8.10). The proof of the existence and uniqueness of the solution of equation (8.10) proceeds just as in §5. We seek a solution in the form of a series

$$(8.11) \qquad y(P) = y_0(P) + \lambda y_1(P) + \lambda^2 y_2(P) + \cdots.$$

As in §5 we find that

$$y_0(P) = f(P), \quad y_{k+1}(P) = \int K(P, Q) y_k(Q) \, dQ, \quad k = 0, 1, 2, \cdots.$$

Applying the lemma proved above, it follows that all the $y_k(P)$ are continuous functions of P. Suppose

$$|f(P)| < N,$$

where N is a constant. Further, let M be the least upper bound of the integral

$$\int |K(P, Q)| \, dQ$$

(M obviously exists). One now sees easily that

$$|y_k(P)| \leq NM^k.$$

It follows that for

$$|\lambda| < (1/M) - \epsilon \quad (\epsilon > 0)$$

the series (8.11) converges uniformly in λ and yields a regular function of λ and a uniformly continuous function of (P, λ). In the same fashion as in §5 it will now be shown that this series gives a solution of the integral equation (8.10), and that no other solution exists in the class of bounded functions.

As in §5 we find that this solution may be represented in the form

$$y(P) = \lambda \int \Gamma(P, Q, \lambda) f(Q) \, dQ + f(P),$$

with

(8.12) $\quad \Gamma(P, Q, \lambda) = K(P, Q) + \lambda K^{(2)}(P, Q) + \lambda^2 K^{(3)}(P, Q) + \cdots.$

The first term of this series is

(8.13) $\qquad\qquad K(P, Q) = \bar{K}(P, Q)/PQ^\alpha, \qquad\qquad \alpha < d,$

where $K(P, Q)$ is a uniformly continuous function of (P, Q). In consequence of the boundedness of G, $\bar{K}(P, Q)$ is also bounded and according to the lemma demonstrated in the beginning of this section,

$$|K^{(2)}(P, Q)| < A/PQ^{2\alpha-d}$$

and, in general,

$$|K^{(m)}(P, Q)| < A_m/PQ^{m\alpha-(m-1)d},$$

if $m\alpha - (m - 1)d > 0$. (See the Remark on p. 31). A and A_m stand for constants. Since $\alpha < d$, for m sufficiently large

$$m\alpha - (m - 1)d < 0.$$

According to the lemma, $K^{(m)}(P, Q)$ is then a uniformly continuous function of (P, Q). All following $K^{(p)}(P, Q)$ are also uniformly continuous. For $p \geq m$

$$|K^{(p+1)}(P, Q)| \leq \left|\int K(P, P_1)K^{(p)}(P_1, Q)\, dP_1\right|$$

$$\leq M_p \int |K(P, P_1)|\, dP_1 \leq M_p M,$$

where M_p is the least upper bound of $|K^{(p)}(P, Q)|$. This yields the proof of the uniform convergence of the series (8.12) in P, Q, λ [for $\lambda < (1/M) - \epsilon$] just as for the series (5.16). The transposed equation may be treated analogously.

All the formulae developed in §5 are equally valid here.

All the considerations of §6 on integral equations with kernels of the form

$$K(P, Q) = \sum_{i=1}^{m} a_i(P)b_i(Q) + K_1(P, Q)$$

are applicable, where $a_i(P)$ and $b_i(Q)$ are continuous in G and $K_1(P, Q)$ has the form (8.13). The proof of the Fredholm theorems follows in the circle

$$|\lambda| < 1/M_1,$$

for M_1 the larger of the upper bounds of the integrals

$$\int |K_1(P, Q)|\, dQ \quad \text{and} \quad \int |K_1(P, Q)|\, dP.$$

In addition it follows that the eigenvalues λ can have no limit point in this circle.

We now turn to the proof of the Fredholm theorems for integral equations with kernels as specified in the section heading. We put

$$\varphi_c(x) = x, \quad \text{if } x \leq c,$$
$$\varphi_c(x) = c, \quad \text{if } x > c.$$

The function

$$K_c(P, Q) = \bar{K}(P, Q)\varphi_c(1/PQ^\alpha)$$

is uniformly continuous in (P, Q) for every c. For sufficiently large c the integrals

$$\int |K(P, Q) - K_c(P, Q)|\, dQ \quad \text{and} \quad \int |K(P, Q) - K_c(P, Q)|\, dP$$

will be arbitrarily small, uniformly in P respectively Q. As we already stated in §6, the uniformly continuous function $K_c(P, Q)$ may be uniformly approximated with arbitrary accuracy in the domain G by sums of the form

$$S_m(P, Q) = \sum_{i=1}^{m} a_i(P)b_i(Q).$$

We then have

$$K(P, Q) = S_m(P, Q) + \tilde{K}(P, Q),$$

where we can make the upper bounds of the values

$$\int |\tilde{K}(P, Q)| \, dQ, \qquad \int |\tilde{K}(P, Q)| \, dP$$

smaller than an arbitrarily small $\epsilon > 0$. The proof for all three Fredholm theorems in the entire λ-plane for integral equations with kernels of the form (8.13) follows from this. In addition, one also obtains the proof that the eigenvalues cannot have a finite limit point.

The proof of the Fredholm theorems for kernels of the form (8.13) carried out above provides basically the proof of these theorems for bounded uniformly continuous kernels. The proof of these latter theorems depends essentially on the fact that certain integrals are small. The requirement of smallness for the integrands was superfluous. We have used this fact in this section.

REMARK. Let the kernel $K(P, Q)$ be a continuous function of P and Q, where $P \in G$, $Q \in G$ and $P \neq Q$, satisfying the condition $|K(P, Q)| < A/PQ^\alpha$, $0 \leq \alpha < d$. Further, let $\epsilon > 0$ and $\alpha + \epsilon < d$. Then

$$K(P, Q) = K(P, Q)PQ^{\alpha+\epsilon}/PQ^{\alpha+\epsilon} = \bar{K}(P, Q)/PQ^{\alpha+\epsilon},$$

where $\bar{K}(P, Q)$ is a continuous function of P and Q. In this manner we have shown that all the Fredholm theorems also hold for kernels of this type.

2. Many problems of mathematical physics lead to integral equations in which the integration is not extended over a region of d-dimensional Euclidean space, but over curves, surfaces, or manifolds of higher dimensions that are imbedded in a Euclidean space of sufficiently high dimension. By a d-dimensional, continuously differentiable manifold M lying in an n-dimensional Euclidean space $E_n (0 < d < n)$, we understand a closed, bounded, and connected point set M in E_n with the property that every point $A \in M$ has a neighborhood in which some $n - d$ coordinates of the points of M are continuously differentiable functions of the remaining d coordinates.

The Fredholm theorems are also valid for integral equations of this sort. Below we shall show, using the considerations of subsection 1, that the Fredholm theorems may be demonstrated in case the integration domain is a closed smooth surface in three-dimensional space. For other manifolds the proof proceeds analogously.

Let us consider the integral equation

$$(8.14) \qquad y(P) = \lambda \int_S K(P, Q) y(Q) \, dS_Q + f(P),$$

where S is a closed smooth surface in three-dimensional space. (We assume that in a sufficiently small neighborhood of an arbitrary point $A \in S$ some one of the coordinates of the points of S is a continuously differentiable function of the other two coordinates). dS_Q is a surface element of S; $P \in S$, $Q \in S$, and $f(P)$ is a given continuous function on S. Further $K(P, Q) = \bar{K}(P, Q)/PQ^\alpha$, where $K(P, Q)$ is continuous for $P, Q \in S$, $0 \leq \alpha < 2$, and PQ is the distance between the points P and Q in three-dimensional space. In order to prove all the Fredholm theorems for equation (8.14) with the help of the considerations of subsection 1, it is enough to show that the Lemma of subsection 1 remains valid and that every continuous kernel $K_1(P, Q)$ given on S may be approximated uniformly with arbitrary accuracy by a degenerate kernel. All other considerations of §§4, 5, 6, 7, and 8 are carried over automatically to the equation under consideration.

The continuous kernel $K_1(P, Q)$ may be considered as a continuous function given in a certain closed set S^2 of six-dimensional space $(x_p, y_p, z_p, x_Q, y_Q, z_Q)$. S^2 is obtained if the points $P(x_p, y_p, z_p)$ and $Q(x_Q, y_Q, z_Q)$ run independently over S. Let R denote a cube in six-dimensional space containing all the points of the set S^2. The continuous functions given on the closed set S^2 may be extended continuously to all of R [7]. According to a theorem of Weierstrass, a continuous function defined on R may be uniformly approximated with arbitrary accuracy by a polynomial. If one now considers this polynomial only on S^2, it appears as a degenerate kernel that approximates $K_1(P, Q)$ with arbitrary accuracy.

We now assure ourselves of the correctness of the Lemma of subsection 1 of this section. To show the continuity of $K_3(P, Q)$ for $P \neq Q$ it is enough to show, in analogy with the proof given in subsection 1, that the integral

$$\int_{S(Q,r)} dS_U/QU^\alpha, \qquad 0 \leq \alpha < 2,$$

taken over the part of the surface S lying in a sphere of radius r centered at Q may be made arbitrarily small uniformly in Q for sufficiently small r. Q varies over a small neighborhood of a certain point Q_0.

Suppose that the surface S is given by a continuously differentiable function $z = f(x, y)$ in the neighborhood of the point Q_0, and Q' and U' are the projections of the points Q and U on the plane $z = 0$. Since $dS < C \, dx \, dy$, with C a constant, and in addition $Q'U' \leq QU$, it follows that

$$\int_{S(Q,r)} dS_U/QU^\alpha < \int_{S(Q',r)} C\ dx\ dy/Q'U'^\alpha.$$

The last integral may be made arbitrarily small for sufficiently small r.

To show the continuity of $K_3(P, Q)$ for $P = Q$ and $\alpha + \beta < d$ it is enough to properly estimate the integral

$$\int_{S(Q,r)+S(P,r)} dS_U/PU^\alpha QU^\beta.$$

P and Q lie in a small neighborhood of the point $P^* = Q^*$ and r tends to zero. We use the inequality (8.8) for this estimate.

In order to prove the correctness of the inequalities (8.3) and (8.4), we show the boundedness of the functions

$$K_3(P, Q)PQ^{\alpha+\beta-d}, \qquad \text{for } \alpha + \beta > d,$$

and

$$K_3(P, Q)/(|\ln PQ| + 1), \qquad \text{for } \alpha + \beta = d,$$

with $P \in S$, $Q \in S$, $P \neq Q$.

To this end we assume that our assertion is false. Then there are sequences of points $P_1 \in S, P_2 \in S, \cdots ; Q_1 \in S, Q_2 \in S, \cdots ; P_i \neq Q_i$ for which

(8.15) $\qquad |K_3(P_i, Q_i)|\, P_i Q_i^{\alpha+\beta-d} \to \infty \qquad$ as $i \to \infty$.

We may assume that the sequences of points P_i and Q_i converge, i.e.

$$P_i \to P_0 \in S, \qquad Q_i \to Q_0 \in S.$$

From the previously demonstrated continuity of the function $K_3(P, Q)$ for $P \neq Q$, it follows that $P_0 = Q_0$. We assume that the z coordinate of the points of S is a continuously differentiable function of x and y in a certain sufficiently small neighborhood $\mathfrak{U}$ of the point P_0, and that the inequality $dS \leq C\ dx\ dy$ (C a constant) holds in this neighborhood. Then we have for all sufficiently large i,

$$|K_3(P_i, Q_i)| \leq A_1 A_2 \int_\mathfrak{U} dS_U/P_i U^\alpha UQ_i^\beta + A_1 A_2 \int_{S-\mathfrak{U}} dS_U/P_i U^\alpha UQ_i^\beta$$

$$\leq A_1 A_2 C \int_{\mathfrak{U}'} dx\ dy/P'_i U'^\alpha U' Q'^\beta_i$$

$$+ A_1 A_2 \max\ \{(1/P_i U^\alpha UQ_i^\beta); U \in S - \mathfrak{U}\} \int_{S-\mathfrak{U}} dS_U,$$

where the prime denotes projection on the plane $z = 0$. Because of the boundedness of the last summand it follows from (8.15) that

(8.16) $$P_i Q_i^{\alpha+\beta-d} \int_{\mathfrak{u}'} dx\, dy / P'_i{}^\alpha U'^\alpha U' Q_i^\beta \to \infty \qquad \text{as } i \to \infty.$$

However for all sufficiently large i we have the inequality $P'_i Q'_i \geq C_1 P_i Q_i$ [$C_1 > 0$ (why?)]. Hence, (8.16) implies that

$$P'_i Q'_i{}^{\alpha+\beta-d} \int_{\mathfrak{u}'} dx\, dy / P'_i{}^\alpha U'^\alpha U' Q'_i{}^\beta \to \infty \qquad \text{as } i \to \infty.$$

But for sufficiently large i, the points P'_i, Q'_i lie in the plane neighborhood $\mathfrak{u}'$ so that the Lemma of subsection 1 gives the result

$$\int_{\mathfrak{u}'} dx\, dy / P'_i{}^\alpha U'^\alpha U' Q'_i{}^\beta < A / P'_i Q'_i{}^{\alpha+\beta-d}$$

(A constant). This, however, contradicts the foregoing. One may show the boundedness of the function $K_3(P, Q)/(|\ln PQ| + 1)$ analogously.

§9. Examples of singular integral equations

A *singular* integral equation is one for which either the Fredholm theorems do not hold or whose eigenvalues have a finite limit point. The singular integral equations introduced in this section have an infinite interval of integration. But if one puts

$$\xi = \tan \eta, \qquad x = \tan y,$$

these integral equations may be transformed into equations with a finite interval of integration.

EXAMPLE 1. The integral equation

$$\varphi(x) = \lambda \int_0^\infty \sin x\xi\, \varphi(\xi)\, d\xi$$

possesses infinitely many linearly independent solutions for $\lambda = \pm(2/\pi)^{\frac{1}{2}}$, since for these values of λ the functions

$$\varphi(x) = (\pi/2)^{\frac{1}{2}} e^{-ax} \pm x/(a^2 + x^2)$$

satisfy the integral equation for arbitrary $a > 0$.

EXAMPLE 2. The equation

$$\varphi(x) = \lambda \int_{-\infty}^\infty e^{-|x-\xi|} \varphi(\xi)\, d\xi$$

has the solution $e^{i\alpha x}$ for $\lambda = (1 + \alpha^2)/2$. Therefore every real $\lambda \geq \frac{1}{2}$ is an eigenvalue. We consider only real values of α since otherwise $e^{i\alpha x}$ is not bounded in the infinite interval $-\infty < x < \infty$.

There are examples of integral equations for which the other Fredholm theorems do not hold.

REFERENCES

1. NYSTRÖM, E. J. *Praktische Auflösung von Intergralgleichungen VII*, Skand. Mat. 170 (1930).
   Über die praktische Auflösung von Intergralgleichungen (Randwertaufagaben), Commentationes Helsingfors 4 (1922), No. 15, 1–25; 5 (1930), No. 5, 1–22.
2. TITCHMARSH, E. C. *The Theory of Functions*, Oxford University Press, 1938, p. 95.
   Knopp, K. *Theory of Functions, Part One*, New York, Dover, 1945, p. 74.
3. KOLMOGOROV, A. N. AND FOMIN, S. V. *Elements of the Theory of Functions and Functional Analysis*, vol. 1: *Metric and Normed Spaces*, Rochester, Graylock, 1957, §14.
4. LOVITT, W. V. *Linear Integral Equations*, McGraw-Hill, New York, 1924, p. 15.
5. COURANT, R. AND HILBERT, D. *Methods of Mathematical Physics*, vol. 1, New York, Interscience, 1953, p. 65.
6. KOWALEWSKI, G. *Intergralgleichungen*, de Gruyter, Berlin and Leipzig, 1930, §13, 179 ff.
7. LEFSCHETZ, S. *Algebraic Topology*, New York, 1942 (American Mathematical Society Colloquium Publications, vol. 27), p. 28.

* Additional references for [1], [4], and [6]: KANTOROVICH, L. V. AND KRYLOV, V. I. *Approximate Methods of Higher Analysis*, P. Noordhoff, Groninger (in preparation).

Chapter II
VOLTERRA INTEGRAL EQUATIONS

§10. Volterra Integral Equations

By Volterra integral equations we understand equations of the form

$$y(P) = \lambda \int K(P, Q) y(Q) \, dQ + f(P),$$

that satisfy the following conditions:

a) Each coordinate of the points P and Q takes on every value between 0 and a constant $a > 0$.

b) $K(P, Q) = 0$ if at least one coordinate of the point Q is greater than the corresponding coordinate (i.e. that having the same index) of the point P.

We shall consider only the one-dimensional case. Then a Volterra integral equation has the form

(10.1) $$y(x) = \lambda \int_0^x K(x, \xi) y(\xi) \, d\xi + f(x).$$

We show that the first Fredholm alternative holds for this equation for every λ, provided that $K(x, \xi)$ is continuous for $0 \leq x \leq a$, $0 \leq \xi \leq x$, and $f(x)$ for $0 \leq x \leq a$.

In other words, we show that *Volterra integral equations have no eigenvalues.*

Proof. The equation (10.1) belongs to the class of integral equations for which we have proved the Fredholm theorems. One has, namely,

$$K(x, \xi) = K(x, \xi) \, |x - \xi|^\epsilon / |x - \xi|^\epsilon, \qquad 0 < \epsilon < 1.$$

The function $\bar{K}(x, \xi)$ defined by

$$\bar{K}(x, \xi) = K(x, \xi) \, |x - \xi|^\epsilon \qquad \text{for } 0 \leq \xi \leq x,$$

$$\bar{K}(x, \xi) = 0 \qquad \text{for } \xi \geq x$$

is uniformly continuous in the square $0 \leq x \leq a$, $0 \leq \xi \leq a$. For this reason, according to §8, all three Fredholm theorems hold for the integral equation (10.1). In order to show that the first Fredholm alternative holds for every λ for this equation, it is enough to show that the corresponding homogeneous equation

(10.2) $$y(x) = \lambda \int_0^x K(x, \xi) y(\xi) \, d\xi$$

has only trivial solutions in the class of continuous functions of x for $0 \leq x \leq a$ and every λ. In order to show this last assertion we denote by B the largest value of $|y(x)|$ for $0 \leq x \leq a$ and by M the largest value of $|K(x, \xi)|$ for $0 \leq x \leq a$, $0 \leq \xi \leq x$. We then obtain from equation (10.2)

$$|y(x)| \leq |\lambda| MBx.$$

Substituting this estimate in the right side of equation (10.2) we further obtain

$$|y(x)| \leq |\lambda|^2 M^2 B x^2 / 2.$$

By continuing this process we obtain the estimates

$$|y(x)| \leq |\lambda|^k M^k x^k B/k! \leq |\lambda|^k M^k a^k B/k!, \qquad k = 1, 2, \cdots.$$

But this last expression tends to zero as $k \to \infty$. Consequently, $y(x) \equiv 0$ on the interval $(0, a)$, which was to be shown.

We now seek a solution of equation (10.1) in the form of a power series

(10.3) $\qquad y(x) = y_0(x) + \lambda y_1(x) + \lambda^2 y_2(x) + \cdots.$

Just as in §8 one must have

$$y_0(x) = f(x), \qquad y_{k+1}(x) = \int_0^x K(x, \xi) y_k(\xi) \, d\xi, \quad k = 0, 1, 2, \cdots.$$

If N is the largest value of $|f(x)|$ in the interval $(0, a)$, we get

$$|y_k(x)| \leq M^k x^k N/k! \leq M^k a^k N/k!.$$

From this it is clear that the series (10.3) is uniformly convergent in λ and x, for λ in an arbitrarily large circle and $0 \leq x \leq a$.

In order to obtain a clear picture of why the Volterra equation has no eigenvalues, we consider (as in §3) the following system of linear algebraic equations that correspond to the Volterra equation in the interval $0 \leq x \leq a$

(10.4) $\qquad f_1 = y_1 - \lambda K_{11} y_1 \Delta \xi$

$\qquad\qquad f_2 = -\lambda K_{21} y_1 \Delta \xi + y_2 - \lambda K_{22} y_2 \Delta \xi$

$\qquad\qquad f_3 = -\lambda K_{31} y_1 \Delta \xi - \lambda K_{32} y_2 \Delta \xi + y_3 - \lambda K_{33} y_3 \Delta \xi,$

$\qquad\qquad \dots\dots\dots\dots\dots\dots\dots\dots\dots\dots\dots\dots\dots$

In the following we keep the notation introduced in §3. The equations (10.4) can be successively solved for arbitrary but fixed λ, provided $|\Delta \xi|$ is sufficiently small, which is what we wish to establish. Indeed, one may solve the first equation for y_1, since the coefficient of y_1 is different from zero for $|\Delta\xi|$ sufficiently small. We substitute this value of y_1 in all subsequent equations. Now one can determine y_2 from the second equation. We substitute this value in all the following equations. Then we determine y_3 from the third equation, and so on. One easily shows that for $\Delta x \to 0$ the solution of the system (10.4) really approaches the solution of the integral equation (10.1).

The determinant of the system (10.4) is

$$D = (1 - \lambda K_{11}\Delta\xi)(1 - \lambda K_{22}\Delta\xi)\cdots(1 - \lambda K_{nn}\Delta\xi),$$

where $\Delta x = \Delta\xi = a/n$, from which it is evident that

(10.5) $$D \geq (1 - |\lambda|M\Delta\xi)^{a/\Delta\xi}.$$

The right hand side of this inequality is different from zero for $\Delta\xi$ sufficiently small, and increases for decreasing $\Delta\xi$. For example, if $\Delta\xi$ is halved, then $(1 - |\lambda|M\Delta\xi)$ in (10.5) is changed to the expression

$$(1 - |\lambda|M\Delta\xi/2)^2 = 1 - |\lambda|M\Delta\xi + \lambda^2 M^2(\Delta\xi)^2/2.$$

As $\Delta\xi \to 0$ the right hand side of (10.5) tends toward

$$e^{-|\lambda|aM}.$$

The algebraic reason for the fact that the Volterra equation has no eigenvalues is that the determinant of the system (10.4) is always different from zero and does not tend to zero as $\Delta\xi \to 0$.

REMARK 1. Completely analogous considerations to those by which we have shown that Volterra equations with uniformly continuous kernels have no eigenvalues can be applied to Volterra equations with kernels of the form

$$K(x, \xi) = \bar{K}(x, \xi)/|x - \xi|^\alpha, \qquad 0 \leq \alpha < 1,$$

where $\bar{K}(x, \xi)$ is a uniformly continuous function.

REMARK 2. We consider the following Volterra equation of the first kind for the unknown function $y(x)$:

(10.6) $$\int_0^x K(x, \xi)y(\xi)\,d\xi = f(x), \qquad f(0) = 0.$$

We assume that $K(x, \xi)$, $K_x(x, \xi)$, $f(x)$ and $f'(x)$ are continuous in $0 \leq x \leq a$ and $0 \leq \xi \leq x$. Every continuous solution $y(x)$ of (10.6) for $0 \leq x \leq a$ satisfies the integral equation

(10.7) $$K(x, x)y(x) + \int_0^x K_x(x, \xi)y(\xi)\,d\xi = f'(x)$$

obtained from (10.6) by termwise differentiation with respect to x. One easily sees that conversely every continuous solution of (10.7) for $0 \leq x \leq a$ satisfies (10.6). If the absolute value of $K(x, x)$ has a positive lower bound, equation (10.7) goes over into a Volterra integral equation of the second kind. If $K(x, x) \equiv 0$, it often is necessary to differentiate equation (10.7) again, and so on.

Chapter III
INTEGRAL EQUATIONS WITH REAL SYMMETRIC KERNELS

§11. Geometric analogies to certain relations between functions (function space)

Let $f(P)$ be uniformly continuous in a given bounded domain G, e.g. in the bounded interval (a, b). Some idea of $f(P)$ is provided by the values of the function on a sufficiently dense point set $P_1, P_2, \cdots, P_n$. In the one-dimensional case one can take for these points,

$$x = a + \Delta x, a + 2\Delta x, \cdots, a + (n-1)\Delta x, a + n\Delta x,$$

where $\Delta x = (b-a)/n$, for n sufficiently large. We denote the values of f at these points by

$$f^{(1)}, f^{(2)}, \cdots, f^{(n)}.$$

We consider these values as the components of the vector $(f^{(1)}, f^{(2)}, \cdots, f^{(n)})$ originating at the origin in n-dimensional Euclidean space. In this way, a vector $(f^{(1)}, f^{(2)}, \cdots, f^{(n)})$ corresponds to the function f.

The length of this vector or its *norm* is equal to

$$[(f^{(1)})^2 + (f^{(2)})^2 + \cdots + (f^{(n)})^2]^{\frac{1}{2}}.$$

Passing to the limit as $n \to \infty$, we call the resultant number $[\int_G f^2(P) \, dP]^{\frac{1}{2}}$ the "length or the *norm of the function* $f(P)$.

In the following table we shall introduce on one side the fundamental quantities and relations that are connected with vectors in n-dimensional Euclidean space, and on the other side the corresponding quantities and relations for functions (in "function space"). In this section *all functions considered will be assumed to be real, defined in a bounded region G, and square integrable* (cf. Remark to §1). The symbol $\int f(P) \, dP$ shall always signify integration over the region G.

1. Vector $f^{(1)}, f^{(2)}, \cdots, f^{(n)}$).
2. Length of a vector:

$$\|f\| = [\sum_{i=1}^{n} (f^{(i)})^2]^{\frac{1}{2}}.$$

3. Distance between the points $(f_1^{(1)}, \cdots, f_1^{(n)})$ and $(f_2^{(1)}, \cdots, f_2^{(n)})$:

$$[\sum_{i=1}^{n} (f_2^{(i)} - f_1^{(i)})^2]^{\frac{1}{2}}.$$

1. Function $f(P)$.
2. Norm of a function:

$$\|f\| = \left[\int f^2(P) \, dP\right]^{\frac{1}{2}}.$$

3. Norm of the difference of two functions $f_2(P) - f_1(P)$:

$$\left[\int (f_2(P) - f_1(P))^2 \, dP\right]^{\frac{1}{2}}.$$

The norm defined above of the difference $f_2(P) - f_1(P)$ characterizes the mean square deviation of the function $f_2(P)$ from $f_1(P)$. This separation between the two functions $f_2(P)$ and $f_1(P)$ may be characterized in many ways, not only by means of the norm defined above of the difference $f_2(P) - f_1(P)$. The separation may be specified, e.g., by the number B, the least upper bound of the expression

$$|f_2(P) - f_1(P)|.$$

If B is small, this signifies that the difference $f_2(P) - f_1(P)$ is uniformly small in the entire region G. Since the region G is bounded, the smallness of the norm of the difference $f_2(P) - f_1(P)$ will follow from the smallness of B. The converse is however false.

If an infinite sequence of functions

$$f_1(P), f_2(P), \cdots, f_k(P), \cdots$$

and a function $f(P)$ are given which in the entire region G satisfy

$$\operatorname{lub}|f(P) - f_k(P)| \to 0, \quad \text{as } k \to \infty,$$

one says that the sequence of functions $f_k(P)$ tends uniformly to $f(P)$.

If however

$$\int [f(P) - f_k(P)]^2 \, dP \to 0 \quad \text{as} \quad k \to \infty,$$

one says that the sequence of functions $f_k(P)$ *converges in the mean* [mean square—Trans.] to $f(P)$. Convergence in the mean follows from uniform convergence for a bounded domain. The following example however shows that the converse of this theorem is false.

The sequence of functions

$$f_k(x) = e^{-kx}$$

converges in the mean to $f(x) \equiv 0$ as $k \to \infty$ for the open interval $(0, 1)$. But this convergence is obviously not uniform.

We give another example of a sequence that converges in the mean but converges nowhere in the ordinary sense. This property is possessed by the sequence of functions $f_i(x)$, $i = 1, 2, 3, \cdots$, defined in the closed interval $(0, 1)$ in the following way:

$$f_{2^k+p}(x) = 1 \quad \text{for } p/2^k \leq x \leq (p+1)/2^k,$$

$$f_{2^k+p}(x) = 0 \quad \text{for } x < p/2^k \text{ and } x > (p+1)/2^k,$$

$$k = 0, 1, 2, \cdots; \quad p = 0, 1, \cdots, 2^k - 1.$$

As $i = 2^k + p \to \infty$ this sequence tends to 0 in the mean. But this sequence does not converge in the ordinary sense for any point of the interval $(0, 1)$, since for every x in the interval there are arbitrarily large i for which $f_i(x) = 1$ and also arbitrarily large i for which $f_i(x) = 0$.

EXERCISE. Show that for a suitable choice of the numbers a_k and b_k the functions

$$|\sin(x - b_k)|^{a_k},\ 1/[1 + a_k(x - b_k)^2],\ e^{a_k(x-b_k)^2}$$

converge in the mean to zero as $k \to \infty$ for the closed interval $(0, 1)$, but do not converge in the ordinary sense at any point.

4. The scalar product of the vectors $(f_1^{(1)}, \cdots, f_1^{(n)})$ and $(f_2^{(1)}, \cdots, f_2^{(n)})$ is defined by the formula

$$\sum_{i=1}^{n} f_1^{(i)} f_2^{(i)}.$$

We shall denote the scalar product by the symbol (f_1, f_2).

4. We define the integral

$$\int f_1(P) f_2(P)\, dP$$

to be the *scalar product of the functions* $f_1(P)$ and $f_2(P)$. The scalar product will be denoted by the symbol (f_1, f_2). The existence of this integral follows from our assumptions and the inequality

$$|ab| \leq \tfrac{1}{2}(a^2 + b^2).$$

5. The triangle inequality (the sum of two sides of a triangle is not smaller than the third side):

$$[\sum(a_i - b_i)^2]^{\frac{1}{2}} + [\sum(b_i - c_i)^2]^{\frac{1}{2}} \geq [\sum(a_i - c_i)^2]^{\frac{1}{2}}.$$

5. The triangle inequality

$$\left[\int (f_1(P) - f_2(P))^2\, dP\right]^{\frac{1}{2}}$$
$$+ \left[\int (f_2(P) - f_3(P))^2\, dP\right]^{\frac{1}{2}}$$
$$\geq \left[\int (f_1(P) - f_3(P))^2\, dP\right]^{\frac{1}{2}}.$$

Both inequalities say exactly the same thing. Accordingly, we shall prove only the first. Without restriction of generality we may set all the b_i equal to zero. Squaring both sides one sees that our inequality is equivalent to the inequality $-\sum a_i c_i \leq [\sum a_i^2 \sum c_i^2]^{\frac{1}{2}}$, since we take all square roots to be non-negative. The last inequality follows immediately from the inequality

(11.1) $$\left(\sum a_i c_i\right)^2 \leq \sum a_i^2 \sum c_i^2,$$

called the Cauchy inequality.

To prove the latter we note that for all real a_i, c_i and λ

$$\sum (a_i \lambda + c_i)^2 \geq 0.$$

§11] GEOMETRIC ANALOGIES (FUNCTION SPACE) 47

Hence the quadratic equation in λ

$$\lambda^2 \sum a_i^2 + 2\lambda \sum a_i c_i + \sum c_i^2 = 0$$

does not have distinct real roots. But this is only possible if inequality (11.1) holds.

The inequality

(11.2) $\qquad \left[\int f(P)\varphi(P)\,dP\right]^2 \leq \int f^2(P)\,dP \cdot \int \varphi^2(P)\,d(P)$

is demonstrated in exactly the same way [1][1].

6. The cosine of the angle between the vectors $(f_1^{(1)}, \cdots, f_1^{(n)})$ and $(f_2^{(1)}, \cdots, f_2^{(n)})$ equals

$$\sum_{i=1}^{n} f_1^{(i)} f_2^{(i)} / \left[\sum_{i=1}^{n} (f_1^{(i)})^2\right]^{\frac{1}{2}} \left[\sum_{i=1}^{n} (f_2^{(i)})^2\right]^{\frac{1}{2}}.$$

According to (11.1) the absolute value of this expression cannot exceed 1. We call a vector of length 1 a *unit vector*.

The cosine of the angle between the unit vectors $(f_1^{(1)}, \cdots, f_1^{(n)})$ and $(f_2^{(1)}, \cdots, f_2^{(n)})$ equals

$$\sum_{i=1}^{n} f_1^{(i)} f_2^{(i)}.$$

7. The condition for orthogonality of the vectors $(f_1^{(1)}, \cdots, f_1^{(n)})$ and $(f_2^{(1)}, \cdots, f_2^{(n)})$ is

$$\sum_{i=1}^{n} f_1^{(i)} f_2^{(i)} = 0.$$

8. The condition for linear dependence (coplanarity) of the vectors $f_k = (f_k^{(1)}, \cdots, f_k^{(n)})$, $k = 1, 2, \cdots, m$, is that there exist constants $c_1, c_2, \cdots, c_m$, not all zero, for which

$$\sum_{k=1}^{m} c_k f_k^{(i)} = 0,$$

for $i = 1, \cdots, n$.

6. We take for the cosine of the angle between the functions $f_1(P)$ and $f_2(P)$ the expression

$$\int f_1(P) f_2(P)\, dP / \left[\int f_1^2(P)\, dP\right]^{\frac{1}{2}} \left[\int f_2^2(P)\, dP\right]^{\frac{1}{2}}.$$

According to (11.2) the absolute value of this expression cannot exceed 1. We call a function $f(P)$ of norm 1 *normed*.

The *cosine of the angle* between the normed functions $f_1(P)$ and $f_2(P)$ equals

$$\int f_1(P) f_2(P)\, dP.$$

7. The conditions for *orthogonality* of the functions $f_1(P)$ and $f_2(P)$ is

$$\int f_1(P) f_2(P)\, dP = 0$$

8. *Linear dependence* of the functions $f_1(P), \cdots, f_m(P)$ is the condition that there exist constants $c_1, \cdots, c_m$, not all zero, so that for all points P the equation

$$\sum_{k=1}^{m} c_k f_k(P) = 0$$

holds.

---
[1] Numbers in brackets refer to the references cited at the end of the chapter.

9. Let m mutually orthogonal unit vectors $\varphi_k = (\varphi_k^{(1)}, \cdots, \varphi_k^{(n)})$, $k = 1, 2, \cdots, m$, be given. The projection $\varphi_k(f)$ of the vector $(f^{(1)}, \cdots, f^{(n)})$ in the direction of the vector $(\varphi_k^{(1)}, \cdots, \varphi_k^{(n)})$ is given by

$$\varphi_k(f) = \sum_{i=1}^{n} f^{(i)} \varphi_k^{(i)}.$$

9. Let m normed mutually orthogonal functions $\varphi_1(P), \cdots, \varphi_m(P)$ be given. We call

$$f_k = \int f(P)\varphi_k(P)\, dP$$

the *Fourier coefficient* of $f(P)$ with respect to the function $\varphi_k(P)$.

THEOREM. *Let $f(P)$ be a square integrable function. We seek constants $c_1, \cdots, c_m$ for which the mean square deviation I_m of the linear combination $c_1\varphi_1(P) + \cdots + c_m\varphi_m(P)$ from the function $f(P)$ is least. The Fourier coefficients f_k yield the minimum value of I_m.* (Give a geometric interpretation of this theorem.)

The proof follows:

$$I_m = \int [f(P) - c_1\varphi_1(P) - \cdots - c_m\varphi_m(P)]^2 \, dP$$

$$= \int f^2(P)\, dP - 2\sum_{k=1}^{m} c_k \int f(P)\varphi_k(P)\, dP$$

$$+ \sum_{i,j=1}^{m} c_i c_j \int \varphi_i(P)\varphi_j(P)\, dP$$

$$= \int f^2(P)\, dP - 2\sum_{k=1}^{m} c_k f_k + \sum_{k=1}^{m} c_k^2$$

$$= \int f^2(P)\, dP + \sum_{k=1}^{m} (f_k - c_k)^2 - \sum_{k=1}^{m} f_k^2.$$

From the above it is clear that I_m assumes its minimum

$$\int f^2(P)\, dP - \sum_{k=1}^{m} f_k^2$$

when $f_k = c_k$, $k = 1, 2, \cdots, m$.

10. For every vector $(f^{(1)}, \cdots, f^{(n)})$ we have the inequality

$$\sum_{k=1}^{m} (\varphi_k(f))^2 \leq \sum_{k=1}^{m} (f^{(k)})^2.$$

In case of equality this relation corresponds to the Pythagorean theorem.

10. For every function $f(P)$ we have the inequality

$$\sum_{k=1}^{m} f_k^2 \leq \int f^2(P)\, dP.$$

(Bessel's inequality.)

Proof of Bessel's inequality. $I_m \geq 0$ is surely true for arbitrary c_i. If we put $c_i = f_i$ for $i = 1, 2, \cdots, m$, then

$$I_m = \int f^2(P)\, dP - \sum_{k=1}^{m} f_k^2, \quad \text{i.e.} \int f^2(P)\, dP \geq \sum_{k=1}^{m} f_k^2.$$

A sequence of normed pairwise orthogonal functions (*orthonormal system*)

(11.3) $\qquad \varphi_1(P), \varphi_2(P), \cdots, \varphi_k(P), \cdots$

is said to be *complete* if for every continuous (and hence also bounded) function $f(P)$ given in a closed bounded region G, the following equation (*Parseval's equation*) holds:

$$\int f^2(P)\, dP = \sum_{k=1}^{\infty} f_k^2.$$

REMARK. By the theorem proved in part 9 of this section we may replace the definition of completeness of a system of functions just given by the following equivalent definition: *The orthonormal system* (11.3) *is called complete, if for an arbitrary continuous function $f(P)$ given in a closed and bounded domain there exists a linear combination $\sum_{k=1}^{m} c_k \varphi_k(P)$ of functions of the system whose mean square deviation from $f(P)$, i.e.*

$$\int [f(P) - \sum_{k=1}^{m} c_k\, \varphi_k(P)]^2\, dP$$

is arbitrarily small. Suppose we define an orthonormal system to be complete if the Parseval equation holds for every square integrable function $f(P)$ that is continuous everywhere in G with the possible exception of finitely many points, curves, and surfaces of dimension up to $(d-1)$ (d is the dimension of the region G on which the functions are defined). This definition is also equivalent to the above condition. This is true because every function $f(P)$ of this class may be approximated in G by a continuous function $f^*(P)$ so that the norm of the difference $f(P) - f^*(P)$ is arbitrarily small. The proof of the equivalence using the triangle inequality (cf. part 5) is left to the reader.

The orthonormal system (11.3) is said to be *closed* if there does not exist a function of the class considered (cf. Remark in §1) such that the integral of its square is positive and which is at the same time orthogonal to all functions (11.3). [In English one conventionally uses the words closed and complete in this connection with meanings interchanged from those given here. Thus the following theorem would read: *Every closed orthonormal system is complete.*—Trans.]

THEOREM. *Every complete orthonormal system is closed.*

Proof. Suppose that the complete system (11.3) is not closed, i.e. there is a function $f(P)$ the integral of whose square is positive and which is orthogonal to all the functions (11.3). For such a function all the Fourier coefficients with respect to the system (11.3) are zero. Consequently the Parseval equation does not hold for $f(P)$.

The converse assertion is false for the class of functions everywhere continuous with the exception of finitely many points, curves, and surfaces of dimension less than or equal to $d-1$. It holds for functions square integrable in the Lebesgue sense (cf. §20, subsection 1).

11. The normal equation of a plane in n-dimensional space

$$(\varphi^{(1)}, \varphi^{(2)}, \cdots, \varphi^{(n)})$$

is $\sum_{i=1}^{n} a^{(i)} \varphi^{(i)} = p,$

where

$$\sum_{i=1}^{n} (a^{(i)})^2 = 1.$$

11. Analogously, we write in function space

$$\int a(P) \varphi(P) \, dP = p,$$

where

$$\int a^2(P) \, dP = 1.$$

12. The equation of a surface of the second order with center at the origin is

(11.4) $\sum_{i,j=1}^{n} K_{ij} \varphi^{(i)} \varphi^{(j)} = 1,$

with $K_{ij} = K_{ji}$.

12. Analogously, we write in function space

(11.5) $\iint K(P,Q) \varphi(P) \varphi(Q) \, dP \, dQ = 1,$

with $K(P, Q) \equiv K(Q, P),$

$$P \in G, Q \in G.$$

13. One of the main theorems in the theory of quadratic forms

(11.6) $\sum_{i,j=1}^{n} K_{ij} \varphi^{(i)} \varphi^{(j)},$

$$K_{ij} = K_{ji},$$

says that it is possible to transform (11.6) by means of a non-singular transformation

(11.7) $\psi^{(i)} = \sum_{j=1}^{n} \varphi_i^{(j)} \varphi^{(j)},$

$$i = 1, 2, \cdots, n,$$

into the canonical form

(11.8) $\sum_{i=1}^{m} (\psi^{(i)})^2 / \lambda_i$

with $m \leq n$. In the sequel we shall be interested only in quadratic forms with real coefficients K_{ij}. In this

13. With enough assumptions on a non-identically vanishing real, symmetric kernel

$$K(P, Q)[K(P, Q) \equiv K(Q, P)]$$

we shall show that the integral form

(11.10) $\iint K(P, Q) \varphi(P) \varphi(Q) \, dP \, dQ$

can be represented in the form of a finite or infinite sum

$$\sum_{i=1} [\psi^i]^2 / \lambda_i$$

(we write no upper limit for i since the sum may be finite or infinite). Here

$$\psi^{(i)} = \int \varphi(P) \varphi_i(P) \, dP$$

case one may choose all the $\varphi_i^{(j)}$ to be real and the λ_i are likewise real. There are many linear transformations (11.7) with real coefficients that transform the quadratic form (11.6) into the canonical form (11.8). Among these the orthogonal transformations, i.e. those for which

$$\sum_{j=1}^{n} \varphi_i^{(j)} \varphi_k^{(j)} = \delta_{ik}$$

play a special role. δ_{ik} is 0 for $i \neq k$ and 1 for $i = k$. It is shown in algebra that the coefficients of the transformation satisfy the equations

(11.9) $$\varphi_i^{(k)} = \lambda_i \sum_{j=1}^{n} K_{kj} \varphi_i^{(j)}$$
$$i = 1, 2, \cdots, m.$$

(Cf. §19. §19 is intended for the reader who wishes to recall the theory of quadratic forms. This theory is exposited there in a form particularly suitable for the sequel. It is recommended that the reader read this section now.)

The transformation of the quadratic form (11.6) by the orthogonal transformation (11.7) into the canonical form (11.8) corresponds geometrically to the adoption of a coordinate system whose axes are the principal axes of the surface (11.4). The vectors $(\varphi_i^{(1)}, \cdots, \varphi_i^{(n)})$, $i = 1, \cdots, m$, are unit vectors in the direction of the bounded principal axes of the surface (11.4). For $m < n$ the surface (11.4) degenerates into a cylindrical surface, having $n - m$ unbounded axes as well as m bounded ones. The semi-axes corresponding to positive λ_i are called the real axes; those corresponding to negative λ_i, the imaginary axes.

and the functions $\varphi_i(P)$, $i = 1, 2, \cdots$, form a non-empty, finite or countably infinite set of orthonormal functions

$$\int \varphi_i(P) \varphi_k(P) \, dP = \delta_{ik}.$$

These functions $\varphi_i(P)$ correspond to the unit vectors in the direction of the bounded principal axes of the surface (11.5). Each of the functions $\varphi_i(P)$ satisfies a homogeneous integral equation

(11.11) $$\varphi_i(P) = \lambda_i \int K(P, Q) \varphi_i(Q) \, dQ.$$

The λ_i are all real. Integral equations with symmetric kernels always have eigenvalues, indeed real eigenvalues (in contrast to Volterra equations).

The problem of finding the unit vector $(\varphi_1^{(1)}, \cdots, \varphi_1^{(n)})$ in the direction of the principal semi-axis of the surface corresponding to λ_1, the λ of smallest absolute value, is equivalent (cf. §19) to the problem of finding the maximum if $\lambda_1 > 0$, or the minimum if $\lambda_1 < 0$, of the form

$$\sum K_{ij}\varphi^{(i)}\varphi^{(j)}$$

subject to the condition

$$\sum_{i=1}^{n} (\varphi^{(i)})^2 = 1.$$

The determination of the unit vector in the direction of the principal semi-axis of the surface (11.4) corresponding to λ_2, or what is the same, solving the system

$$\varphi^{(i)} = \lambda_2 \sum_{j=1}^{n} K_{ij}\varphi^{(j)},$$
$$i = 1, 2, \cdots, n,$$

orthogonal to the solution

$$(\varphi_1^{(1)}, \cdots, \varphi_1^{(n)})$$

is easily reduced to finding the maximum or the minimum of the form

$$\sum_{i,j=1}^{n} (K_{ij} - \varphi_1^{(i)}\varphi_1^{(j)}/\lambda_1)\varphi^{(i)}\varphi^{(j)}$$

in the class of unit vectors

$$(\varphi^{(1)}, \cdots, \varphi^{(n)}).$$

One finds analogously unit vectors in the direction of the other semi-axes (cf. §19).

14. If one substitutes for $\psi^{(i)}$ the given expression in $\varphi^{(j)}$ from (11.7) in the identity

$$\sum_{i,j=1}^{n} K_{ij}\varphi^{(i)}\varphi^{(j)} \equiv \sum_{i=1}^{m} (\psi^{(i)})^2/\lambda_i,$$

The basic idea of the proof of the existence of at least one eigenvalue of the integral equation (11.11) consists in the following (cf. §12). In the class of functions $\varphi(P)$ for which

(11.12) $\quad \int \varphi^2(P)\, dP = 1,$

we show the existence of a function that yields a maximum or minimum λ_1, different from zero, for the integral form (11.10). This function $\varphi_1(P)$ will satisfy the integral equation (11.11) for $i = 1$. The problem of finding the normed function in the direction of the principal axis of the surface (11.5) corresponding to λ_2 and the determination of the corresponding eigenfunction of the integral equation (11.11) is easily reduced to finding the maximum or minimum of the integral form

$$\iint [K(P, Q) - \varphi_1(P)\varphi_1(Q)/\lambda_1]$$

$$\varphi(P)\varphi(Q)\, dP\, dQ$$

for the class of normed functions satisfying condition (11.12). One finds analogously the normed functions directed toward the other semi-axes of the surface (11.5), or what is the same thing, the other normed solutions of the integral equation (11.11) orthogonal to the preceding solutions (cf. subsection 4, §13).

14. Under certain assumptions for $K(P, Q)$ it will be shown in §15 that

$$K(P, Q) = \sum_{i} \varphi_i(P)\varphi_i(Q)/\lambda_i.$$

one obtains

$$\sum_{s,t} K_{st}\varphi^{(s)}\varphi^{(t)} \equiv \sum_{i,j} K_{ij}\varphi^{(i)}\varphi^{(j)}$$
$$\equiv \sum_{i=1}^{m} \sum_{s,t} \varphi_i^{(s)}\varphi^{(s)}\varphi_i^{(t)}\varphi^{(t)}/\lambda_i$$
$$\equiv \sum_{s,t} (\sum_i \varphi_i^{(s)}\varphi_i^{(t)}/\lambda_i)\varphi^{(s)}\varphi^{(t)}.$$

Equating coefficients in the first and last members of this chain of identities one obtains.

$$K_{st} = \sum_{i=1}^{m} \varphi_i^{(s)}\varphi_i^{(t)}/\lambda_i.$$

15. A necessary and sufficient condition that a vector

$$(f^{(1)}, \cdots, f^{(n)})$$

be resolvable in the directions of the finite semi-axes of the surface (11.4) is that the vector $(f^{(1)}, \cdots, f^{(n)})$ be orthogonal to all infinite semi-axes of the surface (11.4). But it easily follows from equation (11.9) that the components $(\chi^{(1)}, \cdots, \chi^{(n)})$ of vectors in the direction of the infinite semi-axes of the surface (11.4) must satisfy the equation

(11.13) $$\sum_{j=1}^{n} K_{ij}\chi^{(j)} = 0,$$
$$i = 1, \cdots, n.$$

The condition that the vector

$$(f^{(1)}, \cdots, f^{(n)})$$

be orthogonal to all vectors

$$(\chi^{(1)}, \cdots, \chi^{(n)})$$

satisfying equation (11.13) is that the system of equations

$$\sum_{j=1}^{n} K_{ji}h^{(j)} = f^{(i)},$$
$$i = 1, \cdots, n,$$

15. In §14 we shall show that every function $f(P)$ that is the *image* of a square integrable function $h(Q)$ by means of the kernel $K(P, Q)$, thus representable in the form

$$f(P) = \int K(P, Q)h(Q)\, dQ,$$

may be developed in a uniformly and absolutely convergent series of eigenfunctions $\varphi_i(P)$ of the kernel $K(P, Q)$ (Hilbert-Schmidt theorem).

or since $K_{ij} = K_{ji}$,

$$\sum_{j=1}^{n} K_{ij} h^{(j)} = f^{(i)},$$

$$i = 1, \cdots, n,$$

have a solution for the unknowns $h^{(1)}, \cdots, h^{(n)}$ (cf. §3).

In §§12–18 we shall assume that all functions considered belong to the class prescribed in the Remark appended to §1, and in addition, that all functions will be real-valued and square integrable in every bounded region where they are defined.

§12. The proof of the existence of eigenfunctions for integral equations with symmetric kernels

1. *Introductory Remarks.* The integral

$$\iint K(P, Q)\varphi(P)\psi(Q) \, dP \, dQ$$

will exist if $\varphi(P)$, $\psi(Q)$, $K(P, Q)$ are square integrable in their domains of definition as was assumed. Indeed,

$$| K(P, Q)\varphi(P)\psi(Q) | \leq \tfrac{1}{2} K^2(P, Q) + \tfrac{1}{2}\varphi^2(P)\psi^2(Q).$$

Therefore

$$\iint | K(P, Q)\varphi(P)\psi(Q) | \, dP \, dQ$$

$$\leq \tfrac{1}{2} \iint K^2(P, Q) \, dP \, dQ + \tfrac{1}{2} \int \varphi^2(P) \, dP \cdot \int \psi^2(Q) \, dQ.$$

In the sequel the symbol $\iint$ will denote integration over the entire domain of definition of $K(P, Q)$, i.e. over the whole domain where $P \in G$, $Q \in G$ (cf. §4 where the symbol is analogously defined).

In §§12–18 we shall consider kernels $K(P, Q)$ that are to be prescribed in part 2 of this section. For such $K(P, Q)$ one may regard all double integrals of the form

$$\iint K(P, Q)\varphi(P)\psi(Q) \, dP \, dQ$$

as repeated single integrals in which one integrates first over Q and then over P [2].

We set

$$B\varphi(P) = \int K(P, Q)\varphi(Q) \, dQ + c\varphi(P)$$

§12] PROOF OF THE EXISTENCE OF EIGENFUNCTIONS

and

$$(\chi, \psi) = \int \chi(P)\psi(P)\, dP.$$

The function $B\varphi(P)$ is square integrable in G, for from the Cauchy-Schwarz inequality

$$\left[\int K(P, Q)\varphi(Q)\, dQ\right]^2 \leq \int [K(P, Q)]^2\, dQ \cdot \int \varphi^2(Q)\, dQ$$

and therefore

$$\int\left[\int K(P, Q)\varphi(Q)\, dQ\right]^2 dP \leq \iint [K(P, Q)]^2\, dP\, dQ \cdot \int \varphi^2(Q)\, dQ.$$

Generalization of the Cauchy-Schwarz inequality. Assume that the sum of the integrals

(12.1) $\quad (B\varphi, \varphi) \equiv \iint K(P, Q)\varphi(P)\varphi(Q)\, dP\, dQ + c \int \varphi^2(P)\, dP,$

c a constant, is *positive definite*. This sum, therefore, is to be non-negative for every real function $\varphi(P)$. We make the blanket assumption that $K(P, Q) = K(Q, P)$. It then follows for arbitrary $\varphi(P)$ and $\psi(P)$ that:

(12.2) $\quad\quad\quad (B\varphi, \psi)^2 \leq (B\varphi, \varphi)(B\psi, \psi).$

Proof of the inequality (12.2). From the assumed positive definiteness of the sum (12.1) it follows for every real μ that

$$\iint K(P, Q)[\varphi(P) + \mu\psi(P)][\varphi(Q) + \mu\psi(Q)]\, dP\, dQ$$
$$+ c \int [\varphi(P) + \mu\psi(P)]^2\, dP \geq 0.$$

Multiplying out we obtain

$$\iint K(P, Q)\varphi(P)\varphi(Q)\, dP\, dQ + \mu \iint K(P, Q)\varphi(P)\psi(Q)\, dP\, dQ$$
$$+ \mu \iint K(P, Q)\varphi(Q)\psi(P)\, dP\, dQ$$
$$+ \mu^2 \iint K(P, Q)\psi(P)\psi(Q)\, dP\, dQ$$
$$+ c \int \varphi^2(P)\, dP + 2\mu c \int \varphi(P)\psi(P)\, dP + c\mu^2 \int \psi^2(P)\, dP \geq 0.$$

Because of the symmetry of $K(P, Q)$ the second and third integrals are equal. One may then write the inequality in the form

$$(B\varphi, \varphi) + 2\mu(B\varphi, \psi) + \mu^2(B\psi, \psi) \geq 0.$$

This last inequality holds for every real μ only if

$$(B\varphi, \psi)^2 \leq (B\varphi, \varphi)(B\psi, \psi).$$

For $K(P, Q) \equiv 0$ and $c = 1$ the inequality (12.1) reduces to the Cauchy-Schwarz inequality

$$\left[\int \varphi(P)\psi(P)\,dP \right]^2 \leq \int \varphi^2(P)\,dP \cdot \int \psi^2(P)\,dP.$$

2. *Until §18 inclusive, we shall consider integral equations with real symmetric kernels of the form*

(12.3) $$K(P, Q) = \bar{K}(P, Q)/PQ^\alpha, \qquad 0 \leq \alpha < d/2,$$

where $\bar{K}(P, Q)$ is a uniformly continuous function in (P, Q) for $P \in G$, $Q \in G$. We shall assume the region G is bounded. The function $\bar{K}(P, Q)$ is therefore bounded.

THEOREM. *Consider a family of functions $h(P)$, for which*

(12.4) $$\int h^2(P)\,dP \leq M^2,$$

where M is a constant, the same for all functions $h(P)$. Then the family of functions $\psi(P)$ in G defined by the equation

$$\psi(P) = \int K(P, Q)h(Q)\,dQ$$

is uniformly bounded and equicontinuous.

We understand by *equicontinuity* of a family of functions the property that for every $\epsilon > 0$ there is an $\eta > 0$ for which $P_1P_2 < \eta$ implies $|\psi(P_2) - \psi(P_1)| < \epsilon$, where η is dependent only on ϵ and not on the function of the family considered and likewise not on the points P_1 and P_2 in G.

Proof.

$$|\psi(P_2) - \psi(P_1)|^2 = \left| \int [K(P_2, Q) - K(P_1, Q)]h(Q)\,dQ \right|^2$$

(12.5) $$\leq \int [K(P_2, Q) - K(P_1Q)]^2\,dQ \cdot \int h^2(Q)\,dQ$$

$$\leq M^2 \int [K(P_2, Q) - K(P_1, Q)]^2\,dQ.$$

The next to last step follows from the Cauchy-Schwarz inequality, the last from the inequality (12.4). We now estimate the integral on the right side of the inequality (12.5). We split the region G into two subregions R and S; R to consist of all points of G whose distance from P_1 or P_2 does not exceed ρ, $S = G - R$. From (12.3) and the boundedness of the function $\bar{K}(P, Q)$ it follows that

$$\int_R [K(P_2, Q) - K(P_1, Q)]^2 \, dQ$$

will be smaller than an arbitrary positive ϵ provided ρ is smaller than a certain $\rho(\epsilon)$ depending only on ϵ and tending to zero with ϵ. $\rho(\epsilon)$ does not depend on the points P_1 and P_2. On the other hand, because of the uniform continuity of $\bar{K}(P, Q)$, the integral

(12.6) $$\int_S [K(P_2, Q) - K(P_1, Q)]^2 \, dQ$$

may be made arbitrarily small for fixed ρ if the points P_1 and P_2 are sufficiently close. The degree of smallness of the integral (12.6) is dependent only on the distance between P_1 and P_2.

We can easily show the uniform boundedness of the family of functions $\psi(P)$ by means of the Cauchy-Schwarz inequality. We have

$$\left| \int K(P, Q) h(Q) \, dQ \right| \leq \left[\int K^2(P, Q) \, dQ \right]^{\frac{1}{2}} \left[\int h^2(Q) \, dQ \right]^{\frac{1}{2}}.$$

The first integral on the right is bounded because of (12.3), the second is less than M from (12.4).

3. THEOREM. *The integral equation*

(12.7) $$\varphi(P) = \lambda \int K(P, Q) \varphi(Q) \, dQ$$

has at least one finite eigenvalue, provided the kernel is not identically zero and has the properties listed at the beginning of the previous subsection.

Generally, we shall consider in the following only kernels with the above properties without specifically mentioning the fact each time.

The idea of the proof presented below is given in the right side of part 13 of §11. Using this method the proof was given almost simultaneously and independently by Hilbert and Holmgren. The greatest difficulty lies in showing that there exists a function $\varphi(P)$ in the class of functions considered satisfying condition (11.12) which maximizes or minimizes the integral form (11.10). The validity of the analogous assertion concerning the existence of a maximum or minimum of the quadratic form (11.6) over the set of unit vectors $(\varphi^{(1)}, \cdots, \varphi^{(n)})$ follows directly from Weierstrass's theorem that every continuous function has a maximum or minimum in an arbitrary closed and bounded set. Thus, in particular, the function (11.6) has one on the sphere $\sum (\varphi^{(i)})^2 = 1$. The proof given here is due to I. M. Gelfand.

Proof. Consider the set S of functions $\varphi(P)$ for which

(12.8) $$\int \varphi^2(P)\, dP = 1.$$

Let

$$I(\varphi) = \iint K(P, Q)\varphi(P)\varphi(Q)\, dP\, dQ.$$

If $\varphi(P)$ belongs to S, then $I(\varphi)$ is bounded. For by the Cauchy-Schwarz inequality,

$$|I(\varphi)|^2 \leq \iint K^2(P, Q)\, dP\, dQ \cdot \iint \varphi^2(P)\varphi^2(Q)\, dP\, dQ$$

$$= \iint K^2(P, Q)\, dP\, dQ \cdot \int \varphi^2(P)\, dP \cdot \int \varphi^2(Q)\, dQ.$$

The first of the last three integrals is finite because of condition (12.3) and the last two are each equal to 1 by (12.8).

Let μ_m, μ_M be the respective lower and upper bounds of the set of values of $I(\varphi)$ on S. We shall show that one at least of the numbers μ_m, μ_M is not zero under the assumption that $K(P, Q)$ is not identically zero. Assuming the contrary, $I(\varphi)$ will be zero for all functions $\varphi(P)$ in S. In particular, it will be zero for every function $\varphi_A(P)$ that is zero everywhere except in a small neighborhood of the point A, where $\varphi_A(P) > 0$. On the other hand, since $K(P, Q)$ is not identically zero, there is a point (A, B) where $K(A, B) \neq 0$. We may assume that B does not coincide with A. For if $K(P, Q)$ were zero for all points (P, Q) for which $P \neq Q$, it would be identically zero because of the assumed uniform continuity in (P, Q) of the function $\bar{K}(P, Q)$ [cf. (12.3)]. But

$$I(\varphi_A + \varphi_B) = I(\varphi_A) + I(\varphi_B) + \iint K(P, Q)\varphi_A(P)\varphi(_B)\, dP\, dQ$$

$$+ \iint K(P, Q)\varphi_A(Q)\varphi_B(P)\, dP\, dQ.$$

The last two integrals are taken over small neighborhoods of the points (A, B) and (B, A). Because of the assumed symmetry of $K(P, Q)$, these two integrals coincide; and since $K(A, B) \neq 0$ and has constant sign in a neighborhood of the point (A, B), these integrals are not zero. The integrals $I(\varphi_A)$ and $I(\varphi_B)$ are zero by assumption. Consequently

$$I(\varphi_A + \varphi_B) \neq 0,$$

which is a contradiction. Thus under our assumptions μ_m and μ_M cannot simultaneously vanish. We suppose $\mu_M \neq 0$.

We now consider an infinite sequence of normed functions

$$\varphi_1(P), \varphi_2(P), \cdots, \varphi_k(P), \cdots,$$

for which

(12.9) $$\lim_{k \to \infty} I(\varphi_k) = \mu_M.$$

One easily shows that the difference

$$-I(\varphi) + \mu_M \int \varphi^2(P)\, dP$$

is positive definite in the sense of part 1 of this section. One may consider in equality (12.2) that

$$B\varphi(P) = \int [-K(P, Q)]\varphi(Q)\, dQ + \mu_M \varphi(P).$$

If we substitute

$$\varphi(P) = \varphi_k(P),$$
$$\psi(P) = B\varphi_k(P),$$

into (12.2), we obtain

(12.10) $$(B\varphi_k, B\varphi_k)^2 \leq (B\varphi_k, \varphi_k)(BB\varphi_k, B\varphi_k).$$

Because of (12.9),

(12.11) $$\lim_{k \to \infty} (B\varphi_k, \varphi_k) = 0.$$

On the other hand, we have

$$|B\varphi_k(P)| \leq \frac{1}{2}\int K^2(P, Q)\, dQ + \frac{1}{2}\int \varphi_k^2(Q)\, dQ + |\mu_M|\,|\varphi_k(P)|.$$

From (12.8) it follows that the first integral is bounded; from (12.8) the second equals 1. Hence,

$$|B\varphi_k(P)| \leq A + |\mu_M|\,|\varphi_k(P)|,$$

where A is a constant independent of φ_k. One obtains an estimate for $BB\varphi_k$ in an analogous manner. Application of the Cauchy-Schwarz inequality shows that $(BB\varphi_k, B\varphi_k)$ is therefore bounded. It then follows from the relations (12.10) and (12.11) that

(12.12) $$\lim_{k\to\infty} (B\varphi_k, B\varphi_k) = 0.$$

Following subsection 2, the family of functions

$$K\varphi_k(P) = \int K(P, Q)\varphi_k(Q)\, dQ$$

is equicontinuous and uniformly bounded. We may then, by Arzelà's theorem [3], choose a uniformly convergent subsequence

$$K\varphi_{k(1)}(P), K\varphi_{k(2)}(P), \cdots, K\varphi_{k(n)}(P), \cdots,$$

from the sequence of functions $K\varphi_k(P)$. We shall show that

$$\lim_{m\to\infty} K\varphi_{k(m)}(P) = \varphi^*(P)$$

is a solution of the integral equation (12.7) for $\lambda = 1/\mu_M$. Indeed,

$$\begin{aligned} |BK\varphi_k(P)| &= |-KK\varphi_k(P) + \mu_M K\varphi_k(P)| \\ &= |K[-K\varphi_k(P) + \mu_M\varphi_k(P)]| \\ &= |KB\varphi_k(P)| \\ &= \left|\int K(P, Q)B\varphi_k(Q)\, dQ\right| \\ &\leq \left\{\int [K(P, Q)]^2\, dQ\right\}^{\frac{1}{2}} \left\{\int [B\varphi_k(Q)]^2\, dQ\right\}^{\frac{1}{2}}. \end{aligned}$$

It now follows from (12.12) that

$$\lim_{m\to\infty} BK\varphi_{k(m)}(P) = 0.$$

Because of the uniform convergence of $K\varphi_{k(m)}(P)$, this implies that

$$B\varphi^*(P) = 0,$$

which we wished to show. $\varphi^*(P)$ is not identically zero, since otherwise the sequence $\mu_M\varphi_{k(m)}(P)$ would tend to zero in the mean for $m \to \infty$ because of the relation (12.12). But this is not possible since $\int [\mu_M\varphi_{k(m)}(P)]^2\, dP = \mu_M^2 > 0$.

REMARKS. 1. The case $\mu_m \neq 0$ is obtained from the case $\mu_M \neq 0$ by changing the sign of $K(P, Q)$.

2. We have shown that the lower (upper) bound of the values of the integral $I(\varphi)$ over the set of normed functions $\varphi(P)$ is equal to the reciprocal of the value of an eigenvalue of the integral equation (12.7), provided only that this bound is different from zero. Conversely, the reciprocal of each eigenvalue λ_i of equation (12.7) is a value of the integral $I(\varphi)$ for a certain function of the class (12.8). Indeed, the value of the integral $I(\varphi)$ for

$\varphi(P) = \varphi_i(P)$, where $\varphi_i(P)$ is the corresponding normed eigenfunction for λ_i, equals

$$\int \varphi_i(P)\, dP \int K(P, Q)\, \varphi_i(Q)\, dQ = (1/\lambda_i) \int \varphi_i^2(P)\, dP = 1/\lambda_i.$$

One may also say that the upper (lower) bound of the values of the integral $I(\varphi)$ for the set of functions satisfying the condition

(12.13) $$\int \varphi^2(P)\, dP \leq 1$$

is equal to the reciprocal of the smallest positive (greatest negative) eigenvalue of equation (12.7), provided this bound is different from zero. It follows that for all functions $\varphi(P)$ satisfying condition (12.13) the integral $I(\varphi)$ does not exceed the reciprocal of the smallest absolute value of the eigenvalues λ of the integral equation (12.7). From the above considerations it is clear that the upper (lower) bound of $I(\varphi)$ over the φ satisfying (12.13) will be attained for a φ, equal to a normed eigenfunction corresponding to the smallest positive (greatest negative) eigenvalue λ, provided that the bound is different from zero.

EXERCISE. Show that the set of values $I(\varphi)$ over the set of φ satisfying (12.8) consists either of a point, a closed, or half-closed interval. The last will occur if $I = 0$ is an endpoint of the interval, in which case this will be the omitted endpoint. The set of values $I(\varphi)$ over the set of φ satisfying (12.13) always includes $I = 0$ and is a point or a closed interval. What cases are possible if the kernel is degenerate?

§13. Some properties of eigenfunctions and eigenvalues for integral equations with symmetric kernels

1. THEOREM. *Eigenfunctions corresponding to distinct eigenvalues of equation* (12.7) *are orthogonal.*

Proof. Let

(13.1) $$\varphi_1(P) = \lambda_1 \int K(P, Q)\varphi_1(Q)\, dQ,$$

(13.2) $$\varphi_2(P) = \lambda_2 \int K(P, Q)\varphi_2(Q)\, dQ,$$

with $\lambda_1 \neq \lambda_2$. Multiply (13.1) by $\lambda_2 \varphi_2(P)$, (13.2) by $\lambda_1 \varphi_1(P)$, subtract termwise and integrate the difference with respect to P. One obtains

(13.3) $$(\lambda_2 - \lambda_1) \int \varphi_1(P)\varphi_2(P)\, dP = \lambda_1 \lambda_2 \iint K(P, Q)\varphi_2(P)\varphi_1(Q)\, dQ\, dP$$

$$- \lambda_1 \lambda_2 \iint K(P, Q)\varphi_1(P)\varphi_2(Q)\, dQ\, dP.$$

If we interchange the variables of integration in the second term on the right hand side, we get

$$\iint K(P, Q)\varphi_1(P)\varphi_2(Q)\, dQ\, dP = \iint K(Q, P)\varphi_2(P)\varphi_1(Q)\, dQ\, dP.$$

Since $K(P, Q) = K(Q, P)$, it follows that the right side of (13.3) is zero. From the assumption $\lambda_1 \neq \lambda_2$ it now follows that

$$\int \varphi_1(P)\varphi_2(P)\, dP = 0,$$

which demonstrates the orthogonality of the two functions.

2. THEOREM. *The eigenvalues of an integral equation with symmetric kernel are all real.*

We shall first show the following lemma.

All eigenfunctions of integral equations of the type under consideration are continuous.

Such functions are to be considered as square integrable according to the remarks at the end of §11. The lemma then follows from subsection 2 of §12.

Proof of the theorem. Suppose the integral equation (12.7) has a complex eigenvalue $\lambda = a + ib$, with $b \neq 0$. Let $\varphi(P)$ be a corresponding eigenfunction. Then

(13.4) $$\varphi(P) = (a + ib) \int K(P, Q)\varphi(Q)\, dQ.$$

If we denote by $\bar{\varphi}(P)$ the complex conjugate function to $\varphi(P)$, we obtain from (13.4)

$$\bar{\varphi}(P) = (a - ib) \int K(P, Q)\bar{\varphi}(Q)\, dQ.$$

According to Theorem 1, we must have

$$\int \varphi(P)\bar{\varphi}(P)\, dP = 0;$$

the lemma then implies $\varphi(P) \equiv 0$. Thus, $a + ib$, with $b \neq 0$, cannot be an eigenvalue.

REMARK. It follows from the above theorem that the real as well as the imaginary part of a complex eigenfunction is also an eigenfunction corresponding to the same eigenvalue.

3. *The orthogonalization of eigenfunctions.* Just as surfaces of the second order may have several principal axes of the same length, so may an integral equation with symmetric kernel have more than one linearly inde-

pendent eigenfunction corresponding to a given eigenvalue. According to the second Fredholm theorem, the set of linearly independent eigenfunctions corresponding to a given eigenvalue is always finite. Suppose the functions are

(13.5) $$\varphi_1(P), \varphi_2(P), \cdots, \varphi_m(P).$$

From the fact that the corresponding eigenvalue is real it follows that we may always take the functions to be real-valued. According to Theorem 1, these functions are all orthogonal to the eigenfunctions corresponding to any other eigenvalue λ of the same integral equation. Also, it is true that a linear combination with constant coefficients of the functions (13.5) is an eigenfunction of the equation (12.7) belonging to the same eigenvalue. We shall show that we can obtain m normed, mutually orthogonal and therefore linearly independent eigenfunctions

$$\psi_1(P), \psi_2(P), \cdots, \psi_m(P),$$

by suitable linear combinations of the functions (13.5). Put

$$\psi_1(P) = a\varphi_1(P).$$

Choose a constant $a \neq 0$ so that

$$\int \psi_1^2(P)\, dP = 1.$$

Set

$$\psi_2(P) = b[\varphi_2(P) + b_1\psi_1(P)],$$

with $b \neq 0$ and b_1 both constant to be determined as follows: choose b_1 so that

$$\int \psi_1(P)\psi_2(P)\, dP = b\left[\int \psi_1\varphi_2\, dP + b_1 \int \psi_1^2(P)\, dP\right] = 0.$$

The fact that $\int \psi_1^2(P)\, dP = 1$ determines b_1 uniquely. The constant b is chosen to make the norm of ψ_2 equal to 1. This is possible since $\varphi_2(P) + b_1\psi_1(P)$ cannot vanish identically because of the assumed linear independence of the functions (13.5). Since all the eigenfunctions of (12.7) are continuous, the integral of the square of $\varphi_2 + b_1\psi_1$ cannot vanish. We further set

$$\psi_3(P) = c[\varphi_3(P) + c_2\psi_2(P) + c_1\psi_1(P)], \qquad c \neq 0.$$

The constant c_1 is chosen so that

$$\int \psi_1\psi_3\, dP = c\left[\int \varphi_3\psi_1\, dP + c_2 \int \psi_2\psi_1\, dP + c_1 \int \psi_1^2\, dP\right] = 0.$$

Since
$$\int \psi_1 \psi_2 \, dP = 0 \quad \text{and} \quad \int \psi_1^2 \, dP = 1,$$
c_1 is uniquely determined to be
$$c_1 = -\int \varphi_3 \psi_1 \, dP.$$
In exactly the same way one chooses c_2 and $c \neq 0$ to yield
$$\int \psi_2 \psi_3 \, dP = 0 \quad \text{and} \quad \int \psi_3^2 \, dP = 1.$$
Continuing this process one obtains $\psi_4(P), \cdots, \psi_m(P)$.

In this way we may restrict ourselves to such linearly independent eigenfunctions of the integral equation which form orthonormal systems. We call an orthonormal system of eigenfunctions of the integral equation *maximal* if every eigenfunction of this integral equation can be expressed as a linear combination of functions of the system. In the following we shall number the eigenfunctions according to the magnitude of the absolute values of the corresponding eigenvalues. (By §8 the set of these eigenvalues has no finite limit point.) We obtain the sequences

(13.6) $\qquad\qquad \varphi_1(P), \varphi_2(P), \cdots, \varphi_i(P), \cdots,$

(13.7) $\qquad\qquad \lambda_1, \lambda_2, \cdots, \lambda_i, \cdots.$

Underneath each eigenfunction $\varphi(P)$ is listed the corresponding eigenvalue λ. The sequences (13.6) and (13.7) may be finite or infinite. The sequence (13.7) may contain equal λ_i, which will then appear consecutively. This will be the case if equation (12.7) has more than one linearly independent eigenfunction belonging to a given eigenvalue λ_i. However, according to the second Fredholm theorem a given λ_i has only a finite number of orthogonal, hence linearly independent, eigenfunctions. If the sequences (13.6) and (13.7) are infinite, it will then follow from this theorem that
$$\lim_{i \to \infty} |\lambda_i| = \infty.$$
The system (13.6) will be maximal in the above sense provided all eigenvalues of the integral equation appear in the sequence (13.7), and the maximum number of mutually orthogonal eigenfunctions belonging to each eigenvalue appear in the sequence (13.6).

4. THEOREM. *Let $\varphi_1(P)$ be the eigenfunction of the integral equation* (12.7) *belonging to the eigenvalue λ_1. Then one obtains for the kernel*
$$K_1(P, Q) = K(P, Q) - \varphi_1(P)\varphi_1(Q)/\lambda_1$$

the sequences of eigenfunctions and eigenvalues, analogous to the sequences (13.6) and (13.7) for the kernel $K(P, Q)$, by deleting $\varphi_1(P)$ and λ_1.

Proof. We shall first show that every eigenfunction $\varphi(P)$ of the kernel $K_1(P, Q)$ corresponding to the eigenvalue λ is an eigenfunction of $K(P, Q)$ belonging to the same eigenvalue. Let

(13.8) $$\varphi(P) = \lambda \int K_1(P, Q)\varphi(Q)\, dQ.$$

Then it follows that

(13.9) $$\int \varphi(P)\varphi_1(P)\, dP = 0,$$

since

$$\int \varphi(P)\varphi_1(P)\, dP = \lambda \iint K_1(P, Q)\varphi(Q)\varphi_1(P)\, dP\, dQ$$

$$= \lambda \iint [K(P, Q) - \varphi_1(P)\varphi_1(Q)/\lambda_1]\varphi_1(P)\varphi(Q)\, dP\, dQ$$

$$= \lambda \iint K(P, Q)\varphi_1(P)\varphi(Q)\, dP\, dQ - (\lambda/\lambda_1) \int \varphi_1(Q)\varphi(Q)\, dQ \int \varphi_1^2(P)\, dP$$

$$= (\lambda/\lambda_1) \int \varphi_1(Q)\varphi(Q)\, dQ - (\lambda/\lambda_1) \int \varphi_1(Q)\varphi(Q)\, dQ = 0.$$

Because of the relation (13.9) the equation (13.8) may be written in the form

$$\varphi(P) = \lambda \int [K(P, Q) - \varphi_1(P)\varphi_1(Q)/\lambda_1]\varphi(Q)\, dQ = \lambda \int K(P, Q)\varphi(Q)\, dQ.$$

In this way we have shown that $\varphi(P)$ is an eigenfunction of the integral equation (12.7) for the same λ. We shall now show, conversely, that every eigenfunction $\varphi_i(P)$ from the sequence (13.6) belonging to an eigenvalue λ_i from the sequence (13.7) for $i > 1$ is also an eigenfunction for the kernel $K_1(P, Q)$ belonging to the same eigenvalue λ_i. Let

(13.10) $$\varphi_i(P) = \lambda_i \int K(P, Q)\varphi_i(Q)\, dQ, \qquad i > 1.$$

Since one now has

$$\int \varphi_1(P)\varphi_i(P)\, dP = 0,$$

it follows from equation (13.10) that

$$\varphi_i(P) = \lambda_i \int [K_1(P, Q) + \varphi_1(P)\varphi_1(Q)/\lambda_1]\varphi_i(Q)\, dQ$$

$$= \lambda_i \int K_1(P, Q)\varphi_i(Q)\, dQ.$$

The function $\varphi_1(Q)$ itself is not an eigenfunction of equation (13.8) since it would otherwise follow from condition (13.9) that $\int \varphi_1^2(P)dP = 0$, which is not possible.

Our theorem follows easily from the assertions proved.

5. If we apply Theorem 4 successively to the kernels

$$K_1(P, Q) = K(P, Q) - \varphi_1(P)\varphi_1(Q)/\lambda_1,$$
$$K_2(P, Q) = K_1(P, Q) - \varphi_2(P)\varphi_2(Q)/\lambda_2,$$
$$\dots\dots\dots\dots\dots\dots\dots\dots\dots\dots\dots\dots$$

we find that all eigenfunctions $\varphi_i(P)$ from the sequence (13.6), belonging to eigenvalues λ_i from the sequence (13.7) for the kernel $K(P, Q)$, are eigenfunctions for the kernels

$$K_m(P, Q) = K(P, Q) - \sum_{k=1}^{m} \varphi_k(P)\varphi_k(Q)/\lambda_k, \qquad i > m,$$

belonging to the same eigenvalues. These eigenfunctions $\varphi_i(P)$, $i > m$, form a maximal system of eigenfunctions for the integral equation with kernel $K_m(P, Q)$ in the sense that every other eigenfunction of this kernel can be represented as a linear combination of the $\varphi_i(P)$, $i > m$.

In this way *one can obtain the sequences (13.6) and (13.7) for a symmetric kernel $K(P, Q)$ by applying the variation method successively to the kernels $K(P, Q)$, $K_1(P, Q)$, $K_2(P, Q)$, $\cdots$*.

6. Suppose the kernel $K(P, Q)$ has only a finite number of linearly independent eigenfunctions (this is the case for degenerate kernels). Then for sufficiently large m the kernel $K_m(P, Q)$ has no eigenvalues. But on the other hand, the functions $\varphi_k(P)$ are continuous by the Lemma of subsection 2, so that the kernel $K_m(P, Q)$ has all the properties prescribed for the kernel $K(P, Q)$ in subsection 2 of §12. Hence, by subsection 3 of §12, $K_m(P, Q) \equiv 0$, i.e.,

(13.11) $\qquad K(P, Q) = \sum_{k=1}^{m} \varphi_k(P)\varphi_k(Q)/\lambda_k.$

It follows from this result that every kernel of the type under consideration that has only finitely many eigenvalues (or, what is the same thing, only finitely many linearly independent eigenfunctions) is degenerate.

REMARK. Let

$$K(P, Q) = \bar{K}(P, Q)/PQ^\alpha, \qquad 0 \leq \alpha < d,$$

where $\bar{K}(P, Q)$ is a uniformly continuous function of P and Q and $\bar{K}(Q, P) = \bar{K}(P, Q)$. It is easy to see that Theorems 1 and 2 of this section remain valid for continuous eigenfunctions of an integral equation with a kernel of this type. Using this result, we shall show that an integral equation with a kernel of this sort has at least one eigenvalue.

It follows from the Lemma of §8 that there is an integer m for which $K^{(m)}(P, Q) = (m\text{-fold}) \, K \circ K \circ \cdots \circ K$ is continuous. Since $K^{(m)}(P, Q)$ is a continuous symmetric kernel, there exists by §12 a real number μ_1 and a continuous function $\varphi_1(P)$ for which

$$\varphi_1 = \mu_1 K^{(m)} \varphi_1$$

[$K^{(m)}$ stands for the operator corresponding to the kernel $K^{(m)}(P, Q)$, cf. p. 24]. We assume that m is odd and put $\mu_1 = \lambda_1^m$, λ_1 real. Further, let e be any primitive mth root of unity. Then, the equation

$$(E - \lambda_1 K)(E - \lambda_1 e K)(E - \lambda_1 e^2 K) \cdots (E - \lambda_1 e^{m-1} K) = (E - \lambda_1^m K^{(m)})$$

holds. The correctness of this equation follows from the algebraic identity

$$(a^m - b^m) \equiv (a - b)(a - eb)(a - e^2 b) \cdots (a - e^{m-1} b).$$

In this way one obtains

$$(E - \lambda_1^m K^{(m)})\varphi_1 = (E - \lambda_1 K)(E - \lambda_1 e K) \cdots (E - \lambda_1 e^{m-1} K)\varphi_1.$$

Set

$$(E - \lambda_1 e K)(E - \lambda_1 e^2 K) \cdots (E - \lambda_1 e^{m-1} K)\varphi_1 = \psi_1.$$

Then we have

$$(E - \lambda_1 K)\psi_1 = 0 \quad \text{and} \quad \psi_1 \not\equiv 0,$$

i.e. ψ_1 is an eigenfunction of the integral equation with kernel $K(P, Q)$. In fact, since m is odd, $\lambda_1 e^p$ is a complex number for $1 \leq p \leq m - 1$. Further, $(E - \lambda_1 e^p K)\varphi \neq 0$ for any $\varphi \neq 0$ and $1 \leq p \leq m - 1$ since according to Theorem 2 of this section an equation with symmetric kernel cannot have any complex eigenvalues. Therefore $\psi_1 \not\equiv 0$, because in the contrary case we would have for some p, $1 \leq p \leq m - 1$ and for a non-identically vanishing function $\varphi(P)$, $(E - \lambda_1 e^p K)\varphi = 0$.

This proves the assertion.

EXERCISE 1. Show that one may find a second eigenfunction from the sequence (13.6) for the symmetric kernel $K(P, Q)$ considered in subsection 2 of §12. Apply for this purpose the variation method given in subsection 3 of §12, with the difference that the admissible functions must now satisfy not only condition (12.8) but also (13.9). Use this method to find the other eigenfunctions.

EXERCISE 2. Let $K(P, Q) = -K(Q, P)$ and also satisfy the condition (12.3). Show that, in this case, every eigenvalue is pure imaginary and that the eigenfunctions cannot be real. An arbitrary eigenfunction is orthogonal to itself and to every other eigenfunction except possibly one belonging to the complex conjugate eigenvalue.

§14. The Hilbert-Schmidt theorem

Every function $f(P)$ that can be represented as the image of a square integrable function $h(P)$, i.e. a function of the form

$$f(P) = \int K(P, Q)h(Q)\,dQ,$$

can be expanded in an absolutely and uniformly convergent series of the eigenfunctions (13.6) of the symmetric kernel $K(P, Q)$.

REMARK. Obviously, it only makes sense to speak of the convergence of this series if the integral equation (12.7) has an infinite number of linearly independent eigenfunctions. Otherwise the series reduces to a finite sum. In order not to lengthen the exposition unduly we shall, as in other analogous cases, always write infinite series, which in the case of only finitely many eigenfunctions will reduce to finite series whose convergence need not be demonstrated.

Proof. The proof will consist in first constructing a certain series of eigenfunctions (13.6), and then showing that it converges uniformly to the function $f(P)$.

Assume that the function $f(P)$ is expanded in a series of eigenfunctions (13.6), these forming an orthonormal system. Let

(14.1) $$f(P) = \sum_{i=1}^{\infty} c_i \varphi_i(P),$$

where the series on the right converges uniformly. To determine the coefficient c_m we multiply both sides of equation (14.1) by $\varphi_m(P)$ and integrate termwise over the whole domain of definition of the functions $f(P)$ and $\varphi_i(P)$. We obtain

$$c_m = \int f(P)\varphi_m(P)\,dP = \iint K(P, Q)h(Q)\varphi_m(P)\,dP\,dQ$$

(14.2) $$= \int h(Q) \left(\int K(P, Q)\varphi_m(P)\,dP \right) dQ$$

$$= \int \varphi_m(Q)h(Q)\,dQ/\lambda_m = h_m/\lambda_m.$$

Here we have put

$$h_m = \int h(Q)\varphi_m(Q)\,dQ.$$

We shall now show that the series

(14.3) $$\sum_{i=1}^{\infty} h_i \varphi_i(P)/\lambda_i$$

is uniformly and absolutely convergent. For this purpose we apply the Cauchy criterion. Considering the section of the series

$$\sum_{i=m}^{m+p} h_i \varphi_i(P)/\lambda_i ,$$

we get from the Cauchy inequality:

(14.4) $$\left[\sum_{i=m}^{m+p} |h_i| \, |\varphi_i(P)/\lambda_i|\right]^2 \leq \sum_{i=m}^{m+p} h_i^2 \sum_{i=m}^{m+p} |\varphi_i(P)/\lambda_i|^2.$$

The coefficients h_i are the Fourier coefficients of the function $h(P)$ with respect to the functions $\varphi_i(P)$. It follows from Bessel's inequality that the series

$$\sum_{i=1}^{\infty} h_i^2$$

converges. Thus from the Cauchy criterion the quantity

$$\sum_{i=m}^{m+p} h_i^2$$

is smaller than an arbitrary positive ϵ_1, if m is sufficiently large.

On the other hand one may regard the quantities

$$\varphi_i(P)/\lambda_i$$

as the Fourier coefficients in the development of the kernel $K(P, Q)$ (considered only as a function of Q) in a series of the functions $\varphi_i(Q)$. Applying Bessel's inequality again, we find

$$\sum_{i=1}^{\infty} (\varphi_i(P)/\lambda_i)^2 \leq \int K^2(P, Q) \, dQ.$$

Because of condition (12.3), the last integral exists and is bounded by a constant independent of P. The sum

$$\sum_{i=m}^{m+p} |\varphi_i(P)/\lambda_i|^2$$

is therefore bounded for arbitrary m and p. We find then from inequality (14.4) that for arbitrary $\epsilon > 0$ there is an m_0, dependent only on ϵ, and for $m > m_0$ and for $p > 0$ we have

$$\sum_{i=m}^{m+p} |h_i \varphi_i(P)/\lambda_i| < \epsilon.$$

Thus the series (14.3) converges absolutely and uniformly according to the Cauchy criterion.

We shall now turn to the proof that the series (14.3) converges to the function $f(P)$.

First we remark that from the theorem proved in subsection 2 of §12 it follows that $f(P)$ and all the eigenfunctions $\varphi_i(P)$ are uniformly continuous.

In addition we have shown above that the series (14.3) is uniformly convergent. It suffices, therefore, in order to show that this series converges to $f(P)$, to demonstrate that the series converges to $f(P)$ in the mean, i.e. that

(14.5) $$\int \left[f(P) - \sum_{i=1}^{m} h_i \varphi_i(P)/\lambda_i \right]^2 dP \to 0,$$

as $m \to \infty$. To prove this last assertion we note that

$$f(P) - \sum_{i=1}^{m} (h_i/\lambda_i)\varphi_i(P) = \int \left[K(P, Q) - \sum_{i=1}^{m} \varphi_i(P)\varphi_i(Q)/\lambda_i \right] h(Q) \, dQ.$$

To abbreviate we put

$$K_m(P, Q) = K(P, Q) - \sum_{i=1}^{m} \varphi_i(P)\varphi_i(Q)/\lambda_i,$$

$$g_m(P) = f(P) - \sum_{i=1}^{m} (h_i/\lambda_i)\varphi_i(P).$$

We now have

$$\int \left[f(P) - \sum_{i=1}^{m} (h_i/\lambda_i)\varphi_i(P) \right]^2 dP$$

$$= \iint \left[K(P, Q) - \sum_{i=1}^{m} \varphi_i(P)\varphi_i(Q)/\lambda_i \right] h(Q) \left[f(P) - \sum_{i=1}^{m} (h_i/\lambda_i)\varphi_i(P) \right] dP \, dQ$$

(14.6) $$= \frac{1}{2} \iint K_m(P, Q)[h(Q) + g_m(Q)][h(P) + g_m(P)] \, dP \, dQ$$

$$- \frac{1}{2} \iint K_m(P, Q)h(Q)h(P) \, dP \, dQ - \frac{1}{2} \iint K_m(P, Q)g_m(Q)g_m(P) \, dP \, dQ.$$

We have made use of the equation

$$\iint K_m(P, Q)h(P)g_m(Q) \, dP \, dQ = \iint K_m(P, Q)h(Q)g_m(P) \, dP \, dQ,$$

which holds because of the symmetry of the kernel $K(P, Q)$. Because

$$\int g_m^2(P) \, dP = \int f^2(P) \, dP - \sum_{i=1}^{m} (h_i/\lambda_i)^2 \leq \int f^2(P) \, dP,$$

there exists a number M independent of m for which

$$\int g_m^2(P) \, dP < M, \quad \int h^2(P) \, dP < M,$$

$$\int [h(P) + g_m(P)]^2 \, dP < M$$

for every m. It follows from Remark 2 of subsection 3 of §12 and from subsection 5 of §13 that the estimate

$$\left| \int K_m(P, Q)\psi(P)\psi(Q) \, dP \, dQ \right| \leq M/|\lambda_{m+1}|$$

holds for functions $\psi(P)$ satisfying

$$\int \psi^2(P) \, dP \leq M.$$

Hence all three integrals on the right side of (14.6) tend to zero as $m \to \infty$, which proves the relation (14.5).

CONSEQUENCE. *The Schmidt formula for the solution of an integral equation with symmetric kernel.*

Consider the integral equation

(14.7) $$\varphi(P) = \lambda \int K(P, Q)\varphi(Q) \, dQ + f(P).$$

$K(P, Q)$ is a symmetric kernel of the kind specified at the beginning of subsection 2 of §12, $f(P)$ a known uniformly continuous function, $\varphi(P)$ the unknown function, and λ a parameter. According to the first Fredholm theorem (§8), the integral equation (14.7) has a uniformly continuous solution $\varphi(P)$ for exactly those λ that are not eigenvalues. According to the Hilbert-Schmidt theorem the function $\varphi(P) - f(P)$ can be expanded in an absolutely and uniformly convergent series of the eigenfunctions of the kernel $K(P, Q)$. Let

$$\varphi(P) - f(P) = \sum_{i=1}^{\infty} c_i \varphi_i(P).$$

Hence

(14.8) $$\varphi(P) = \sum_{i=1}^{\infty} c_i \varphi_i(P) + f(P).$$

If we substitute the expression on the right hand side of equation (14.8) for $\varphi(P)$ into equation (14.7), we obtain

(14.9) $$\sum_{i=1}^{\infty} c_i \varphi_i(P) + f(P) = \lambda \sum_{i=1}^{\infty} c_i \int K(P, Q)\varphi_i(Q) \, dQ$$
$$+ \lambda \int K(P, Q)f(Q) \, dQ + f(P)$$
$$= \lambda \sum_{i=1}^{\infty} c_i \varphi_i(P)/\lambda_i + \lambda \sum_{i=1}^{\infty} (f_i/\lambda_i)\varphi_i(P) + f(P).$$

We have made use of the Hilbert-Schmidt theorem and replaced

$$\int K(P, Q)f(Q) \, dQ$$

by the absolutely and uniformly convergent series

$$\sum_{i=1}^{\infty} (f_i/\lambda_i)\varphi_i(P), \quad \text{with} \quad f_i = \int f(P)\varphi_i(P)\,dP.$$

It is also easy to see that the first series on the right hand side of (14.9) converges uniformly. Comparing coefficients of $\varphi_i(P)$ on the two sides of equation (14.9), we get

$$c_i = \lambda(c_i/\lambda_i) + \lambda(f_i/\lambda_i).$$

Hence

$$c_i = \lambda f_i/(\lambda_i - \lambda)$$

and consequently

(14.10) $\qquad \varphi(P) = \lambda \sum_{i=1}^{\infty} f_i\varphi_i(P)/(\lambda_i - \lambda) + f(P),$

where the series on the right converges uniformly and absolutely. This equation is named after E. Schmidt.

In case λ coincides with an eigenvalue λ_i, one can obtain the solution of equation (14.7) in an analogous manner. Now one has $f_i = 0$ for all i corresponding to the eigenvalue λ in virtue of the third Fredholm theorem. The solution may be written in the form of series (14.8), in which $c_i = \lambda f_i/(\lambda_i - \lambda)$ for $\lambda_i \neq \lambda$ and $c_i = \alpha_i$ for $\lambda_i = \lambda$, where α_i is an arbitrary constant.

EXERCISE. Prove the Hilbert-Schmidt theorem for degenerate kernels directly from formula (13.11).

§15. A theorem on the expansion of the kernel

THEOREM. *The kernel $K(P, Q)$ considered in this chapter can be expanded in a series*

(15.1) $\qquad \sum_{i=1}^{\infty} \varphi_i(P)\varphi_i(Q)/\lambda_i$

that converges in the mean to $K(P, Q)$ with respect to P, i.e. for each fixed Q:

(15.2) $\qquad \lim_{m\to\infty} \int \left[K(P,Q) - \sum_{i=1}^{m} \varphi_i(P)\varphi_i(Q)/\lambda_i \right]^2 dP = 0.$

Proof. We consider the function

$$K_2(P, Q) = \int K(P, S)K(S, Q)\,dS$$

as dependent only on P, holding Q fixed. By the Hilbert-Schmidt theorem we may expand this function in a series of the eigenfunctions $\varphi_i(P)$, which

converges absolutely and uniformly in P. Let
$$K_2(P, Q) = \sum_{i=1}^{\infty} c_i \varphi_i(P).$$
From (14.2) follows that
$$c_i = \int K(P, Q) \varphi_i(P)/\lambda_i \, dP = \varphi_i(Q)/\lambda_i^2.$$
Consequently

(15.3) $$K_2(P, Q) = \sum_{i=1}^{\infty} \varphi_i(P) \varphi_i(Q)/\lambda_i^2.$$

This series converges according to the Hilbert-Schmidt theorem absolutely and uniformly in P for each fixed Q. From the symmetry of the kernel follows now the absolute and uniform convergence of the series in Q for each fixed P. One may not yet quite conclude that the series converges uniformly in P and Q. We shall prove such convergence in §17.

From (15.3) follows

(15.4) $$K_2(Q, Q) = \sum_{i=1}^{\infty} \varphi_i^2(Q)/\lambda_i^2.$$

This series converges but we have not yet shown that it converges uniformly in Q.

On the other hand

(15.5)
$$\int \left[K(P, Q) - \sum_{i=1}^{m} \varphi_i(P)\varphi_i(Q)/\lambda_i \right]^2 dP$$
$$= \int K(P, Q)K(Q, P) \, dP - 2 \sum_{i=1}^{m} (\varphi_i(Q)/\lambda_i) \int K(Q, P)\varphi_i(P) \, dP$$
$$+ \sum_{i=1}^{m} \varphi_i^2(Q)/\lambda_i^2$$
$$= K_2(Q, Q) - 2 \sum_{i=1}^{m} \varphi_i^2(Q)/\lambda_i^2 + \sum_{i=1}^{m} \varphi_i^2(Q)/\lambda_i^2$$
$$= K_2(Q, Q) - \sum_{i=1}^{m} \varphi_i^2(Q)/\lambda_i^2.$$

From (15.4), the last difference tends to zero as $m \to \infty$. Hence (15.2) follows, as was to be shown.

§16. Classification of kernels

We consider the integral form

(16.1) $$\iint K(P, Q)\chi(P)\chi(Q) \, dP \, dQ,$$

where $\chi(P)$ is any square integrable function. By means of the Cauchy-Schwarz inequality one may easily show that the integral (16.1) exists, since $K(P, Q)$ is square integrable [cf. (12.3)]. By the Hilbert-Schmidt theorem one may expand the function of P

$$\int K(P, Q)\chi(Q)\, dQ$$

in a series of eigenfunctions $\varphi_i(P)$ of the kernel $K(P, Q)$, uniformly convergent in P. Using (14.2) we obtain

(16.2) $$\int K(P, Q)\chi(Q)\, dQ = \sum_{i=1}^{\infty} (x_i/\lambda_i)\varphi_i(P)$$

with

$$x_i = \int \chi(P)\varphi_i(P)\, dP.$$

After we multiply both sides of (16.2) by $\chi(P)$ and integrate over P, we obtain

(16.3) $$\iint K(P, Q)\chi(P)\chi(Q)\, dP\, dQ = \sum_{i=1}^{\infty} x_i^2/\lambda_i.$$

The integral form and the kernel $K(P, Q)$ are called *positive (negative) definite* if the integral form (16.1) is non-negative (non-positive) for every square integrable function $\chi(P)$ (cf. §12, subsection 1).

It is obvious from equation (16.3) that *a necessary and sufficient condition that the form* (16.1) *be positive (negative) definite is that all the λ_i are positive (negative)*.

We shall call the form (16.1) and the kernel $K(P, Q)$ *positive (negative) quasidefinite* if all the eigenvalues with the possible exception of finitely many are positive (negative).

§17. Dini's theorem and its applications

DINI'S THEOREM. *If a monotone sequence of continuous functions*

(17.1) $$f_1(P), f_2(P), \cdots, f_k(P), \cdots$$

converges everywhere in a closed and bounded set F to a continuous function $f(P)$, then the sequence converges uniformly.

Proof. Without loss of generality we can assume $f(P) \equiv 0$, since subtracting $f(P)$ from each function in the sequence (17.1) will reduce the general case to the special one. We may also assume that the sequence (17.1) is monotone decreasing at each point P (that means $f_k(P) \geq f_{k+1}(P)$ for $k = 1, 2, 3, \cdots$). If it is not, we need only change the sign of every $f_k(P)$.

§17] DINI'S THEOREM AND ITS APPLICATIONS 75

Then, let there be given a sequence of continuous functions, monotone decreasing at each point P, convergent to zero at each point of a closed and bounded set F. We shall show that the convergence is uniform. To this end we note the following. For every $\epsilon > 0$ and every point P of the set F one may determine an m so that

$$0 \leq f_m(P) < \epsilon.$$

Because of the continuity of $f_m(P)$, this inequality will also hold in a neighborhood O_P of the point P. And since the sequence under consideration decreases monotonically in O_P, it follows that

(17.2) $\qquad 0 \leq f_k(P) < \epsilon \qquad$ for all $k \geq m$.

One can, therefore, for each ϵ determine a neighborhood O_P for every point P of the set F in which the inequality (17.2) holds from some m on. By the Heine-Borel theorem one may select a finite set from the family of neighborhoods O_P that will cover F. Let M be the maximum of the corresponding m. It follows that

$$f_k(P) < \epsilon$$

for the whole set F and all $k \geq M$, which proves the theorem.

APPLICATIONS OF DINI'S THEOREM.

1. In §15 we showed that

(15.4) $\qquad K_2(Q, Q) = \sum_{i=1}^{\infty} \varphi_i^2(Q)/\lambda_i^2.$

At that time we left open the question of the uniform convergence of the series in Q. Now this question is easily answered using Dini's theorem. By the Lemma in §8, the function $K_2(Q, Q)$ is uniformly continuous in Q. Consequently we may assume the function continuous in a domain $\bar{G}$ (cf. remark on p. §§§). The sequence

$$\sum_{i=1}^{m} \varphi_i^2(Q)/\lambda_i^2$$

forms for each point Q a monotone sequence of continuous functions converging to a continuous function as $m \to \infty$. By Dini's theorem the convergence is *uniform in* $\bar{G}$.

2. Since the series on the right side of (15.4) converges uniformly in Q, the sequence (15.2) converges to zero uniformly in Q [cf. (15.5)]. Therefore we have

$$\lim_{m \to \infty} \iint \left[K(P, Q) - \sum_{i=1}^{m} \varphi_i(P)\varphi_i(Q)/\lambda_i \right]^2 dP\, dQ = 0.$$

3. The uniform convergence of the series (15.4) in Q implies the *absolute and uniform convergence of the series* (15.3) *in* (P, Q), since

$$\sum_{i=m}^{m+p} |\varphi_i(P)\varphi_i(Q)/\lambda_i^2| \leq \tfrac{1}{2} \sum_{i=m}^{m+p} \varphi_i^2(P)/\lambda_i^2 + \tfrac{1}{2} \sum_{i=m}^{m+p} \varphi_i^2(Q)/\lambda_i^2.$$

4. On integration of both sides of (15.4) with respect to Q we find that

$$\int K_2(Q, Q)\, dQ = \sum_{i=1}^{\infty} 1/\lambda_i^2,$$

since the functions $\varphi_i(Q)$ are normed. Thus we have shown that *the series of reciprocal squares of the λ_i is convergent.*

5. MERCER'S THEOREM. *If the kernel $K(P, Q)$ is quasidefinite and uniformly continuous in (P, Q), then the series (15.1) converges to $K(P, Q)$ not only in the mean but absolutely and uniformly in (P, Q).*

Proof. We assume that the kernel is positive quasidefinite. Further we remark that $K_m(P, P) \geq 0$ if the kernel $K_m(P, Q)$ is positive definite and continuous. Indeed, if $K_m(A, A) < 0$ at a point A, then the kernel will be negative in a neighborhood in the (P, Q)-space of the point (A, A) due to continuity. We construct, then, a continuous function $\varphi_A(P)$ zero everywhere in G with the exception of a certain small neighborhood G_A of the point A, where it is positive. If then G_A is sufficiently small, we have

$$\iint K_m(P, Q)\varphi_A(P)\varphi_A(Q)\, dP\, dQ < 0,$$

which contradicts the assumption of positive definiteness of the kernel $K_m(P, Q)$.

Because of the assumed positive quasidefiniteness of the kernel $K(P, Q)$, the kernel

$$K_m(P, Q) = K(P, Q) - \sum_{i=1}^{m} \varphi_i(P)\varphi_i(Q)/\lambda_i$$

will be positive definite for m sufficiently large (cf. §16). From the fact shown above for positive definite kernels it follows that for all sufficiently large m

$$K(P, P) \geq \sum_{i=1}^{m} \varphi_i^2(P)/\lambda_i.$$

Consequently the series

(17.3) $$\sum_{i=1}^{\infty} \varphi_i^2(P)/\lambda_i$$

converges for all P, and therefore the series

(17.4) $$\sum_{i=1}^{\infty} \varphi_i^2(Q)/\lambda_i$$

converges for all Q. Since $K(P, P)$ is bounded, the partial sums of the series (17.3) and (17.4) are also bounded for all P and Q. Let $M_0 > 0$ be a common upper bound for the absolute values of the partial sums.

We apply the Cauchy inequality, choosing m so large that for $i \geq m$ all the $\lambda_i > 0$. We obtain

(17.5) $$\left[\sum_{i=m}^{m+p} |\varphi_i(P)\varphi_i(Q)/\lambda_i|\right]^2 \leq \sum_{i=m}^{m+p} \varphi_i^2(P)/\lambda_i \sum_{i=m}^{m+p} \varphi_i^2(Q)/\lambda_i.$$

If we apply the Cauchy criterion to the series (17.4), we find that for arbitrary fixed Q there corresponds to every $\epsilon > 0$ an m_0, dependent only on ϵ and Q, for which

$$\sum_{i=m}^{m+p} \varphi_i^2(Q)/|\lambda_i| = \sum_{i=m}^{m+p} \varphi_i^2(Q)/\lambda_i < \epsilon^2/2M_0$$

if $m > m_0(\epsilon, Q)$. Therefore, by (17.5), for arbitrary $p > 0$ and $m > m_0(\epsilon, Q)$,

$$\sum_{i=m}^{m+p} |\varphi_i(P)\varphi_i(Q)/\lambda_i| < \epsilon.$$

Hence, by the Cauchy criterion, the series (15.1) converges uniformly and absolutely in P for fixed Q.

Because of the relation (15.2) demonstrated above, it follows that the series (15.1) converges to $K(P, Q)$. In particular,

(17.6) $$K(P, P) = \sum_{i=1}^{\infty} \varphi_i^2(P)/\lambda_i.$$

Since $K(P, P)$ is continuous in the closed domain $\overline{G}$ by assumption, and all functions $\varphi_i(P)$ are continuous in the same domain (cf. Lemma in subsection 2, §12), and all λ_i are positive from a certain i onward, the series on the right side of (17.6) converges uniformly in P according to Dini's theorem. Therefore for every $\epsilon > 0$ there is an m_0, dependent only on ϵ so that for arbitrary $p > 0$

$$\sum_{i=m}^{m+p} \varphi_i^2(P)/\lambda_i < \epsilon \qquad \text{for } m > m_0(\epsilon).$$

From inequality (17.5) it further follows that for arbitrary P and Q, $p > 0$ and the same m and ϵ

$$\sum_{i=m}^{m+p} |\varphi_i(P)\varphi_i(Q)/\lambda_i| < \epsilon.$$

Therefore the series (15.1) converges absolutely and uniformly in (P, Q), which proves our assertion.

§18. Example

Consider the integral equation

(18.1) $$\varphi(x) = \lambda \int_0^l G(x, \xi)\varphi(\xi)\, d\xi, \qquad 0 \leq x \leq l,$$

where $G(x, \xi)$ is the Green's function constructed in §2. As was shown, this function is symmetric in its two variables. We may therefore apply the entire theory developed in this chapter to the integral equation (18.1). The equation has, as we have seen in §2, infinitely many eigenfunctions and eigenvalues, all of which can be found as in §2. We norm the eigenfunctions, write the corresponding eigenvalues under each eigenfunction to obtain the sequences

(18.2) $$\begin{cases} (2/\pi)^{\frac{1}{2}} \sin x, & (2/\pi)^{\frac{1}{2}} \sin 2x, \quad \cdots, \quad (2/\pi)^{\frac{1}{2}} \sin kx, \quad \cdots \\ \quad 1, & \quad 4, \quad \cdots, \quad\quad\quad\quad k^2, \quad \cdots. \end{cases}$$

To simplify, we have put in equation (2.1) $l = \pi$ and $c = \rho/T_0 = 1$. In this case only one eigenfunction corresponds to each eigenvalue. The eigenfunctions belonging to different eigenvalues are easily verified to be orthogonal to one another in agreement with the general theory (cf. subsection 1, §13).

Applying the Hilbert-Schmidt theorem, we find that every function $f(x)$ of the form

$$(18.3) \qquad f(x) = \int_0^\pi G(x, \xi) h(\xi) \, d\xi, \qquad 0 \leq x \leq 1,$$

where $h(\xi)$ is a square integrable function, can be expanded in a series of eigenfunctions (18.2) of the kernel $G(x, \xi)$. We assume, as we have already previously done everywhere (cf. Remark in §1), that the function $h(\xi)$ has only a finite number of discontinuities. We differentiate both sides of equation (18.3) twice with respect to x making use again of formula (2.1) as in §2. It then follows that, with the possible exception of finitely many points, the equation

$$f''(x) = -h(x)/T_0$$

holds. Conversely, using formula (2.1), one may show that every function $f(x)$ continuously differentiable in the closed interval $(0, \pi)$, vanishing at the endpoints, with a second derivative continuous with the exception of finitely many points and square integrable, can be represented in the form (18.3), where $h(x)$ is square integrable. One takes $-f''(x)T_0$ for $h(x)$. It then follows from the Hilbert-Schmidt theorem that every function $f(x)$ continuously differentiable in the closed interval $(0, \pi)$, satisfying $f(0) = f(\pi) = 0$, and having a square integrable second derivative continuous except at a finite number of points, can be expanded in an absolutely and uniformly convergent series of the functions $\sin kx$.

It is known from the theory of trigonometric series that it is possible to obtain such an expansion with weaker assumptions on $f(x)$. It is, for example, sufficient for $f(x)$ to be continuously differentiable in the closed interval $(0, \pi)$ and to have $f(0) = f(\pi) = 0$[4]. From this last class of functions one can easily choose a function that does not satisfy the requirement of the Hilbert-Schmidt theorem. It is sufficient, e.g., to pick a function that has no second derivative anywhere. This shows that the conditions of the Hilbert-Schmidt theorem for the possibility of expanding the function $f(x)$ in an absolutely and uniformly convergent series in the eigenfunctions is not necessary.

We now show that the system (18.2) is complete, i.e. in the closed interval $(0, \pi)$ one may find for every continuous function $f(x)$ a linear combination of the $\sin kx$ whose mean square deviation from $f(x)$ is arbitrarily small. To this end we remark that in the closed interval $(0, \pi)$ one can find for every continuous function $f(x)$ a function $f_1(x)$ with continuous second derivative, vanishing at the ends, such that the norm of the difference $f(x) - f_1(x)$ is arbitrarily small. As we have shown above, the function $f_1(x)$ can be expanded in a uniformly convergent series in $\sin kx$. Thus one can approximate the function $f(x)$ by a linear combination of the $\sin kx$ (namely by a partial sum of the series in $\sin kx$ that converges uniformly to this function) for which the mean square error is arbitrarily small. It follows, making use of the triangle inequality (cf. subsection 5, §11), that the system (18.2) is complete.

The closure of the system (18.2) follows from its completeness as in subsection 10, §11.

Since all the eigenvalues of the kernel $G(x, \xi)$ are positive, Mercer's theorem is applicable. We obtain

$$(\pi/2)G(x, \xi) = \sin x \sin \xi/1 + \sin 2x \sin 2\xi/4 + \cdots,$$

where the series on the right is uniformly and absolutely convergent in (x, ξ).

REFERENCES

1. The inequality (11.1) is first found in Cauchy (1821); Oeuvres, II, T. III (1897), p. 373. Cauchy derived it from the identity

$$\sum_{i=1}^{n} a_i^2 \cdot \sum_{i=1}^{n} b_i^2 - \left(\sum_{i=1}^{n} a_i b_i\right)^2 = \frac{1}{2} \sum_{i=1}^{n} \sum_{j=1}^{n} (a_i b_j - a_j b_i)^2$$

using the fact that the right side cannot be negative. The inequality (11.2) was first demonstrated and systematically used by Bunyakovskiĭ [*Sur quelques inégalités concernant les intégrales ordinaires*, Mémoires de l'Acad. de Ste. Petersbourg (VII) 1 (1859), No. 9]. In the literature this inequality is often named after H. A. Schwarz, even though he first found it in 1885 [Werke, I (1890), p. 251]. TRANSLATORS' REMARK: Following the English usage, we shall term (11.2) the Cauchy-Schwarz inequality.
2. TAYLOR, A. E. *Advanced Calculus*, Ginn & Co., 1955, p. 529.
3. KOLMOGOROV AND FOMIN (see Ref. 3 of Ch. I), §17 (Arzelàs theorem). COURANT-HILBERT (see Ref. 5, Ch. I), p. 59.
4. COURANT-HILBERT, p. 70.

APPENDIX

§19. Reduction of a quadratic form to canonical form by means of an orthogonal transformation

We shall give here the proof of the possibility of such a reduction, which corresponds to the proofs for the theory of integral forms in §§12 and 13.

1. First we present certain properties of orthogonal unit vectors. Let

$$\varphi_k = (\varphi_k^{(1)}, \cdots, \varphi_k^{(n)}), \qquad k = 1, \cdots, n,$$

be n such unit vectors, i.e., the relations

$$\sum_{i=1}^{n} \varphi_k^{(i)} \varphi_p^{(i)} = \delta_{kp}, \qquad k, p = 1, \cdots, n,$$

where $\delta_{kp} = 0$ for $k \neq p$ and $\delta_{pp} = 1$, hold.

a) $D = |\varphi_p^{(i)}| = \pm 1$. Indeed, if one computes D^2 by the well known rule as a product of two equal determinants, one obtains a determinant whose principal diagonal consists of ones and all other elements equal zero.

b) Let $\Phi_k^{(i)}$ be the cofactor of the element $\varphi_k^{(i)}$ in the determinant D. Then we have

(19.1) $$\Phi_k^{(i)} = D \varphi_k^{(i)}.$$

The equation

$$\sum_{i=1}^{n} (\Phi_k^{(i)} - D\varphi_k^{(i)}) \varphi_p^{(i)} = 0, \qquad p = 1, \cdots, n,$$

holds for arbitrary k. Since D is not zero, one obtains the relation (19.1) if one considers the above equations as a linear homogeneous system of n equations with the coefficients $\varphi_p^{(i)}$.

c) Multiplying equation (19.1) by $\varphi_k^{(j)}$ and summing over k, we obtain

(19.2) $$\sum_{k=1}^{n} \varphi_k^{(i)} \varphi_k^{(j)} = \delta_{ij}.$$

2. Consider the values of the quadratic form

(19.3) $$\sum_{i,j=1}^{n} K_{ij} \varphi^{(i)} \varphi^{(j)}, \qquad K_{ij} = K_{ji},$$

on the sphere

(19.4) $$\sum_{i=1}^{n} (\varphi^{(i)})^2 = 1,$$

with all the K_{ij} and $\varphi^{(i)}$ real.

By Weierstrass's theorem there exists at least one point in the closed and bounded point set, the sphere (19.4), where the continuous function (19.3) assumes its greatest value. Let this greatest value be μ_1, assumed at the point $A_1(\varphi_1^{(1)}, \cdots, \varphi_1^{(n)})$. [See Remark 2 to subsection 3, §12, where the existence of a function $\varphi(P)$ on the sphere

$$\int \varphi^2(P)\, dP = 1$$

is shown for which the integral form

$$\iint K(P,\, Q)\varphi(P)\varphi(Q)\, dP\, dQ$$

assumes its greatest value, provided this is different from zero.]

We now consider the values of the form (19.3) at the points of the intersection S_{n-2} of the sphere (19.4) with the hyperplane perpendicular to the vector $(\varphi_1^{(1)}, \cdots, \varphi_1^{(n)})$ and going through the center of the sphere. By the same theorem, there is at least one point in S_{n-2} for which the form (19.3) assumes a maximum relative to the points of S_{n-2}. Let this value be μ_2 assumed at the point $A_2(\varphi_2^{(1)}, \cdots, \varphi_2^{(n)})$.

We consider further the values of the form (19.3) at the points of the set S_{n-3}, the intersection of S_{n-2} with the hyperplane perpendicular to the vector $(\varphi_2^{(1)}, \cdots, \varphi_2^{(n)})$ and going through the origin. Let the upper bound of the values of (19.3) on S_{n-3} be μ_3, assumed at the point $A_3(\varphi_3^{(1)}, \cdots, \varphi_3^{(n)})$.

If we continue this procedure, we shall obtain n perpendicular unit vectors $\varphi_k = (\varphi_k^{(1)}, \cdots, \varphi_k^{(n)})$, $k = 1, \cdots, n$. We assume these to be the directions of new coordinate axes $O\psi_1, \cdots, O\psi_n$ and obtain

$$\psi^{(k)} = \sum_{i=1}^{n} \varphi_k^{(i)} \varphi^{(i)}, \qquad k = 1, \cdots, n.$$

Each of the sets S_{n-k} will be the intersection of an $(n - k + 1)$-dimensional plane

$$\psi^{(1)} = \psi^{(2)} = \psi^{(3)} = \cdots = \psi^{(k-1)} = 0$$

and the sphere

(19.5) $$\sum_{i=1}^{n} (\psi^{(i)})^2 = 1.$$

That the sphere (19.4) transforms into the sphere (19.5) and hence equation (19.5) is satisfied for all the points of S_{n-k} follows from

$$\sum_{k=1}^{n} (\psi^{(k)})^2 = \sum_{k=1}^{n} \sum_{i,j=1}^{n} \varphi_k^{(i)} \varphi^{(i)} \varphi_k^{(j)} \varphi^{(j)}$$
$$= \sum_{i,j=1}^{n} \left(\sum_{k=1}^{n} \varphi_k^{(i)} \varphi_k^{(j)} \right) \varphi^{(i)} \varphi^{(j)}$$

and from (19.2)

$$\sum_{k=1}^{n} \varphi_k^{(i)} \varphi_k^{(j)} = \delta_{ij}.$$

In the new coordinates $\psi^{(i)}$ the form (19.3) assumes the form

(19.6) $$F \equiv \sum_{i,j=1}^{n} K_{ij} \varphi^{(i)} \varphi^{(j)} \equiv \sum_{i,j=1}^{n} K^{*}{}_{ij} \psi^{(i)} \psi^{(j)}$$
$$\equiv \sum_{i=1}^{n} \mu_i (\psi^{(i)})^2.$$

The fact that $K^{*}{}_{ii} = \mu_i$ follows from the fact that the form assumes the value μ_i at the point A_i where all the coordinates of ψ are zero except $\psi^{(i)}$, which is 1. The assertion $K^{*}{}_{1j} = K^{*}{}_{j1} = 0$ for $j > 1$ can be proved as follows. Assume that for a certain $j > 1$, $K^{*}{}_{1j} \ne 0$. Set all $\psi^{(i)}$ except $\psi^{(1)}$ and $\psi^{(j)}$ equal to zero. We then have

$$F = \mu_1 (\psi^{(1)})^2 + 2K^{*}{}_{1j} \psi^{(1)} \psi^{(j)} + \mu_j (\psi^{(j)})^2.$$

Let $|\psi^{(j)}|$ be very small in comparison with $|\psi^{(1)}|$ and

$$(\psi^{(1)})^2 + (\psi^{(j)})^2 = 1.$$

We may then neglect the terms of order $(\psi^{(j)})^2$ and obtain

(19.7) $$F \approx \mu_1 + 2K^{*}{}_{1j} \psi^{(j)}.$$

We now choose the sign of $\psi^{(j)}$ so that $2K^{*}{}_{1j} \psi^{(j)} > 0$. It then follows from the relation (19.7) that there is a point on the sphere (19.5) or, what is the same thing, on the sphere (19.4) for which $F > \mu_1$, which contradicts the definition of μ_1. We have therefore shown that $K^{*}{}_{1j} = 0$ for $j > 1$. The proof that all other $K^{*}{}_{ij}$ are zero for $i \ne j$ proceeds analogously.

We now have the sequence of numbers

$$\mu_1 \geq \mu_2 \geq \cdots \geq \mu_n.$$

It is possible that one of these numbers is zero and the following are non-positive. By renumbering the μ_i and the corresponding ψ_i we can arrange to have

$$\mu_1 \geq \cdots \geq \mu_i \geq \cdots \geq \mu_m; \quad \mu_{m+1} = \cdots = \mu_n = 0 \quad (\mu_i \ne 0 \text{ for } i \leq m).$$

If we put

$$\lambda_i = 1/\mu_i, \qquad i = 1, \cdots, m,$$

then equation (19.6) will take the form

$$\sum_{i,j=1}^{n} K_{ij} \varphi^{(i)} \varphi^{(j)} = \sum_{i=1}^{m} (\psi^{(i)})^2 / \lambda_i.$$

If $\lambda_1, \cdots, \lambda_\nu$ are positive and $\lambda_{\nu+1}, \cdots, \lambda_m$ negative, $\lambda_1^{\frac{1}{2}}, \lambda_2^{\frac{1}{2}}, \cdots, \lambda_\nu^{\frac{1}{2}}$ are called the real semi-axes of the surface

(19.8) $$\sum_{i,j=1}^{n} K_{ij} \varphi^{(i)} \varphi^{(j)} \equiv \sum_{i=1}^{m} (\psi^{(i)})^2 / \lambda_i = 1.$$

This surface has the intercepts $\pm \lambda_i^{\frac{1}{2}}$ on the axis $O\psi^{(i)}$ for $i = 1, 2, \cdots, \nu$;

$\lambda_1^{\frac{1}{2}}$ is the smallest real semi-axis. The numbers $(-\lambda_{\nu+1})^{\frac{1}{2}}, \cdots, (-\lambda_m)^{\frac{1}{2}}$ are called the *imaginary* semi-axes of the surface (19.8); the latter does not meet the real axes $O\psi^{(\nu+1)}, \cdots, O\psi^{(m)}$.

If $m < n$, the equation (19.8) represents a cylindrical surface in the space $(\varphi^{(1)}, \cdots, \varphi^{(m)})$, as well as in the space $(\psi^{(1)}, \cdots, \psi^{(m)})$ whose generating plane is

$$\psi^{(1)} = \cdots = \psi^{(m)} = 0.$$

In this case we say that the semi-axes corresponding to the axes $O\psi^{(m+1)}, \cdots, O\psi^{(n)}$ are infinite.

In the following subsections we shall revert to the previous numbering of the axes.

3. We shall now show that

(19.9) $\qquad \mu_1 \varphi_1^{(i)} = \sum_{j=1}^{n} K_{ij} \varphi_1^{(j)}, \qquad i = 1, 2, \cdots, n.$

According to our considerations we must have for every real $\varphi^{(j)}$

$$F \equiv \mu_1 \sum_{i=1}^{n} (\varphi^{(i)})^2 - \sum_{i,j=1}^{n} K_{ij} \varphi^{(i)} \varphi^{(j)} \geq 0.$$

If $\varphi^{(i)} = \varphi_1^{(i)}$ for every i, then $F = 0$, i.e. it assumes its minimum value. The partial derivatives with respect to each variable must vanish for these values of $\varphi^{(i)}$. This gives (19.9).

4. In order to find the axis $O\psi^{(2)}$, instead of considering the values of the form (19.3) on the set S_{n-2} (the intersection of the sphere (19.4) and the hyperplane $\psi^{(1)} = 0$) one may, if $\mu_2 > 0$, investigate the values of the form

(19.10) $\qquad \sum_{i,j=1}^{n} (K_{ij} - \mu_1 \varphi_1^{(i)} \varphi_1^{(j)}) \varphi^{(i)} \varphi^{(j)}$

on the whole sphere (19.4). One can easily show that the form (19.10) assumes its greatest value μ^* at a certain point $A^*(\varphi_*^{(1)}, \cdots, \varphi_*^{(n)})$ belonging to S_{n-2} and that $\mu^* = \mu_2$. We have, in fact,

(19.11) $\qquad \sum_{i,j=1}^{n} [K_{ij} - \mu_1 \varphi_1^{(i)} \varphi_1^{(j)}] \varphi^{(i)} \varphi^{(j)}$
$\qquad = \sum_{i,j=1}^{n} K_{ij} \varphi^{(i)} \varphi^{(j)} - \mu_1 [\psi^{(1)}]^2 = \sum_{i=2}^{n} \mu_i [\psi^{(i)}]^2.$

Therefore, in order that the form (19.10) assume its greatest value at the point A^* on the sphere (19.4) or (19.5), it is necessary that for this point all the $\psi^{(i)}$ vanish with the exception of those i for which $\mu_i = \mu_2$ if $\mu_2 > 0$. The sum of the squares of these last must equal 1. At this point we maintain the original enumeration of the μ_i for which

$$\mu_2 \geq \mu_3 \geq \cdots \geq \mu_n;$$

μ^* must consequently be equal to μ_2. The point A^* lies on S_{n-2} and we rename it A_2. This method for finding the second semi-axis is also applicable in the case when $\mu_2 \leq 0$, since then the form (19.11) assumes zero as its greatest value, e.g. for

$$\psi_1 = 1, \quad \psi_2 = \cdots = \psi_n = 0.$$

In exactly the same way, one can reduce the search for the axis $O\psi_3$ for $\mu_3 > 0$ to the question of where on the sphere (19.4) the form

(19.12) $$\sum_{i,j=1}^{n} [K_{ij} - \mu_1 \varphi_1^{(i)} \varphi_1^{(j)} - \mu_2 \varphi_2^{(i)} \varphi_2^{(j)}] \varphi^{(i)} \varphi^{(j)}$$

has its maximum, etc.

If one applies the same considerations as in subsection 3 to the form (19.10), one may show that $\varphi_2^{(i)}$, $i = 1, 2, \cdots, n$, must satisfy the equation

$$\mu_2 \varphi_2^{(i)} = \sum_{j=1}^{n} [K_{ij} - \mu_1 \varphi_1^{(i)} \varphi_1^{(j)}] \varphi_2^{(j)}$$

or

(19.13) $$\mu_2 \varphi_2^{(i)} = \sum_{j=1}^{n} K_{ij} \varphi_2^{(j)}, \quad i = 1, \cdots, n,$$

for $\mu_2 > 0$ because

$$\sum_{j=1}^{n} \varphi_1^{(j)} \varphi_2^{(j)} = 0.$$

If we use the form (19.12) and the other analogously constructed forms, we can show that the values $\mu_3 > 0$, $\mu_4 > 0$, $\cdots$ with corresponding $\varphi_3^{(i)}$, $\varphi_4^{(i)}$, $\cdots$ also satisfy equations of the form (19.9).

However, if $\mu_2 = 0$, then the quadratic form

$$\sum_{i,j=1}^{n} [K_{ij} - \mu_1 \varphi_1^{(i)} \varphi_1^{(j)}] \varphi^{(i)} \varphi^{(j)} \leq 0$$

for arbitrary $\varphi^{(i)}$, $i = 1, 2, \cdots, n$, and also vanishes if $\varphi^{(i)} = \varphi_2^{(i)}$ for all i. If we put the partial derivatives of this form equal to zero for $\varphi^{(i)} = \varphi_2^{(i)}$, then we obtain equations of the form (19.9). The same considerations can be applied to the other vectors $\varphi_k^{(i)}$ which correspond to $\mu_k = 0$.

If $\mu_2 < 0$ then the numbers $\mu_2, \mu_3, \cdots, \mu_n$ are all negative and the corresponding vectors $(\varphi_k^{(1)}, \cdots, \varphi_k^{(n)})$, $k = 2, \cdots, n$, are found by looking for the minimum (instead of maximum) of the form (19.11). The numbers $\mu_2, \mu_3, \cdots, \mu_n$ and the corresponding vectors are obtained in reverse order. One may proceed analogously in case $\mu_3 < 0$, etc. In all cases equation (13.19) and its analogues for the vectors $(\varphi_k^{(1)}, \cdots, \varphi_k^{(n)})$ still hold.

EXERCISE. Based on equations (19.9), (19.13), $\cdots$ work out a non-variational method of reducing the quadratic form to canonical form using the solutions of the characteristic equation

$$|K_{ij} - \mu \delta_{ij}| = 0.$$

§20. Theory of integral equations with symmetric kernels that are square integrable in the Lebesgue sense

The theory of integral equations carried out above may be easily carried over to integral equations whose kernels are symmetric and square integrable in the Lebesgue sense, strengthening the theory.

The development of the new theory agrees in the main with that carried out in §§11–16 and is effected by the same plan. We shall, therefore, confine ourselves in the development to those places in the theory that differ materially from the foregoing theory.

1. We shall assume that the reader is acquainted with the theory of Lebesgue integration [1][1]. We note briefly some properties of Lebesgue integrable functions (summable functions).

a) FUBINI'S THEOREM. Suppose the function $f(P, Q)$ is integrable over the topological product of two measurable sets G and H, where $P \in G$ and $Q \in H$. We shall write this integral in the form

$$I = \int_G \int_H f(P, Q) \, dP \, dQ.$$

Then, the integral

$$I(Q) = \int_G f(P, Q) \, dP$$

exists for almost all points Q in H. The function $I(Q)$ is summable and

(20.1) $$I = \int_H I(Q) \, dQ.$$

Conversely, if the integral

$$I^*(Q) = \int_G |f(P, Q)| \, dP$$

exists for almost all points $Q \in H$, if the function $I^*(Q)$ is summable over H, and if $f(P, Q)$ is measurable on GH, then the integral I exists and equation (20.1) holds.

We recall that the totality of points (P, Q) with $P \in G$ and $Q \in H$ is called the topological product GH of the sets G and H. We assume that the sets G, H, and GH are subsets of the Euclidean spaces $(x_1, \cdots, x_d)$, $(y_1, \cdots, y_e)$ and $(x_1, \cdots, x_d, y_1, \cdots, y_e)$. If the point P respectively Q is defined by the coordinates $(x_1, \cdots, x_d)$ respectively $(y_1, \cdots, y_e)$ then the point (P, Q) is determined by the coordinates $(x_1, \cdots, x_d, y_1, \cdots, y_e)$. The measure of the sets G, H, respectively GH is taken to be the Lebesgue measure in the Euclidean spaces of dimension d, e, respectively $d + e$.

---

[1] Numbers in brackets refer to the references cited at the end of the section.

b) Let the function $f(S)$ be defined in a d-dimensional region G. Suppose further that the integral of $f(S)$ over every d-dimensional cube in the interior of G with edges parallel to the coordinate axes is zero. Then $f(S) = 0$ almost everywhere in G.

c) RIESZ-FISCHER THEOREM. Suppose

$$f_1(P), f_2(P), \cdots, f_n(P), \cdots$$

is an infinite sequence of square integrable functions on a measurable set G. Suppose further that for every $\epsilon > 0$ there is a number N such that

$$(20.2) \qquad \int_G (f_n - f_m)^2 \, dP < \epsilon$$

if $n > N$ and $m > N$. Then, there exists in G a function $f(P)$ that is square integrable and which satisfies

$$(20.3) \qquad \lim_{n \to \infty} \int_G (f_n - f)^2 \, dP = 0.$$

The converse assertion that (20.2) follows from (20.3) is obvious.

In the sequel we shall always understand the convergence of a sequence of functions to be in the mean. The Riesz-Fischer theorem is the analogue to the well known necessary and sufficient Cauchy criterion of convergence. A function space in which the Cauchy criterion of convergence for sequences of functions holds is called *complete*. In other words, *the space of square summable functions, with convergence understood to be mean convergence, is complete*.

d) For the class of functions that are square summable, one may carry out all the considerations and prove all the theorems that were shown in subsections 1–10 of §11 for functions that had discontinuities only at finitely many points, curves, and k-dimensional surfaces ($k = 2, 3, \cdots, d - 1$), and for which integration was understood to be in the Riemann sense. In addition one may show that for this class the concepts of completeness and closure of an orthonormal system are equivalent. The proof given in subsection 10, §11 that the closure of a system follows from its completeness may be carried over in toto to the class of square summable functions that we are considering. One can show as follows that *for this class the completeness of the system follows from its closure*.

Suppose that the orthonormal system of functions

$$(20.4) \qquad \varphi_1(P), \quad \varphi_2(P), \cdots, \quad \varphi_k(P), \cdots$$

were not complete. Then there would be a function $f(P)$, square summable, for which

(20.5) $$\int f^2(P)\, dP - \sum_{k=1}^{\infty} a_k^2 > 0,$$

with

$$a_k = \int f(P)\varphi_k(P)\, dP.$$

We now consider the sequence of partial sums of the series

(20.6) $$\sum_k a_k \varphi_k(P).$$

For this sequence the Cauchy convergence criterion is fullfilled, because

$$\int \left[\sum_{k=m+1}^{m+n} a_k \varphi_k(P)\right]^2 dP = \sum_{k=m+1}^{m+n} a_k^2,$$

and the infinite series $\sum_k a_k^2$ converges. According to the Riesz-Fischer theorem, there must exist a square summable function $\varphi(P)$ to which the series (20.6) converges in the mean. It will then follow that the function

$$f(P) - \varphi(P)$$

is orthogonal to all the functions $\varphi_k(P)$ and that the integral of its square is equal to the left side of (20.5) and hence is positive. Consequently, the system (20.4) is not closed.

e) We record one more important property of square summable functions, that we shall use later.

For every function $F(P)$ that is square summable over a measurable set G, and for every $\epsilon > 0$ there is a continuous function $f(P)$ with the property

$$\int [F(P) - f(P)]^2\, dP < \epsilon\ [2].$$

2. For Lebesgue integration the values of the integrand on a set of measure zero have no influence on the value of the integral. One may even leave this function undefined on a set of measure zero. We may, therefore, regard all functions that differ only on a set of measure zero, on which they need not even be defined, as equivalent. Accordingly, we shall term a solution of the integral equation any function $\varphi(P)$ that is square summable and satisfies the equation

(20.7) $$\varphi(P) = \lambda \int_G K(P, Q)\varphi(Q)\, dQ$$

for almost all P.

The proof that the integral equation (20.7) with a real symmetric kernel has a non-trivial solution in the class of square summable functions for a certain λ follows the same plan as in §12. Hence, we shall only carry out the part of the proof that differs materially from the corresponding part in §12. We shall assume that the integral

$$(20.8) \qquad \iint K^2(P, Q) \, dP \, dQ$$

extended over the topological product of two equal measurable sets G to which P and Q belong, exists. It follows from Fubini's theorem that the integral

$$\int K^2(P, Q) \, dQ$$

exists for almost all P, and consequently that the integral

$$\int K(P, Q)\varphi(Q) \, dQ$$

exists for almost all P for every square summable function $\varphi(Q)$, since

$$|K(P, Q)\varphi(Q)| \leq \tfrac{1}{2}[K^2(P, Q) + \varphi^2(Q)].$$

Here and in the sequel the symbol $\iint$ will denote integration over the topological product of two copies of the measurable set G to which the points P and Q belong; the symbol $\int$ will denote integration over G.

For every function $\varphi(P)$ with a summable square the integral

$$\iint K(P, Q)\varphi(P)\varphi(Q) \, dP \, dQ$$

exists, since

$$|K(P, Q)\varphi(P)\varphi(Q)| \leq \tfrac{1}{2}[K^2(P, Q) + \varphi^2(P)\varphi^2(Q)]$$

and

$$\iint \varphi^2(P)\varphi^2(Q) \, dP \, dQ = \int \varphi^2(P) \, dP \cdot \int \varphi^2(Q) \, dQ.$$

Henceforth, we shall only consider symmetric kernels $K(P, Q)$ for which the integral (20.8) exists. In general, *we shall only consider functions whose square is Lebesgue integrable over the whole domain of definition G*, without stating this explicitly each time.

3. Subsection 1, §12 remains entirely the same. Instead of the theorem proved in subsection 2, §12 we show the following. *Let H be a family of*

functions $h(P)$ each of which satisfies

(20.9) $$\int h^2(P)\, dP \leq M^2, \qquad M > 0.$$

M *is a constant, the same for all functions $h(P)$. Then the family Σ of functions $\psi(P)$ defined by the equation*

$$\psi(P) = \int K(P, Q) h(Q)\, dQ$$

is compact, i.e. from every infinite sequence of such functions one may always extract a sequence that converges in the mean.

Proof. If $K(P, Q)$ for $P \in G$ and $Q \in G$ is a uniformly continuous function, then by the theorem proved in subsection 2, §12 the family is equicontinuous and uniformly bounded.

By Arzelà's theorem the family Σ is compact, i.e. from every infinite sequence of functions $\psi(P)$ one can choose a subsequence that converges uniformly and hence in the mean. Using this fact, we shall prove the compactness of the family Σ for an arbitrary kernel $K(P, Q)$ with summable square.

Following e) of subsection 1 of this section, we can construct a sequence of continuous functions

$$K_1(P, Q), \quad K_2(P, Q), \cdots, K_n(P, Q), \cdots$$

for which

(20.10) $$\iint [K(P, Q) - K_n(P, Q)]^2\, dP\, dQ < 1/2^{2n}, \quad n = 1, 2, \cdots.$$

Let

(20.11) $$\psi_1(P), \quad \psi_2(P), \cdots, \quad \psi_n(P), \cdots$$

be an infinite sequence of functions from the family Σ obtained from the sequence of functions

(20.12) $$h_1(P), \quad h_2(P), \cdots, h_n(P), \cdots$$

of the family H by means of the equations $\psi_i = K h_i$ (cf. (6.2)). Since $K_1(P, Q)$ is continuous, one may extract from the sequence (20.12) an infinite subsequence

(20.13) $$h_1^{(1)}(P), \quad h_2^{(1)}(P), \cdots, \quad h_n^{(1)}(P), \cdots$$

for which the sequence of functions

(20.14) $$K_1 h_1^{(1)}, \quad K_1 h_2^{(1)}, \cdots, \quad K_1 h_n^{(1)}, \cdots$$

converges uniformly and hence in the mean. Further, we can extract from the sequence (20.13) a new infinite subsequence

(20.15) $\qquad h_1^{(2)}(P), \qquad h_2^{(2)}(P), \cdots, \qquad h_n^{(2)}(P), \cdots$

for which the sequence

$$K_2 h_1^{(2)}, \qquad K_2 h_2^{(2)}, \cdots, \qquad K_2 h_n^{(2)}, \cdots$$

converges in the mean, etc.

It is easily seen that the sequence

(20.16) $\qquad h_1^{(1)}(P), \qquad h_2^{(2)}(P), \cdots, \qquad h_n^{(n)}(P), \cdots$

is so constructed that the sequence

(20.17) $\qquad K_m h_1^{(1)}, \qquad K_m h_2^{(2)}, \cdots, \qquad K_m h_n^{(n)}, \cdots$

converges in the mean for arbitrary $K_m(P, Q)$.

We shall now show that the sequence of functions $\psi(P)$ of the family Σ

$$K h_1^{(1)}, \qquad K h_2^{(2)}, \cdots, \qquad K h_n^{(n)}, \cdots,$$

which is a subsequence of (20.11), likewise converges in the mean. By the triangle inequality (cf. §11, subsection 5) we have

(20.18) $\quad \begin{aligned} \| K h_m^{(m)} - K h_n^{(n)} \| &\leq \| K h_m^{(m)} - K_p h_m^{(m)} \| \\ &\quad + \| K_p h_m^{(m)} - K_p h_n^{(n)} \| + \| K_p h_n^{(n)} - K h_n^{(n)} \|. \end{aligned}$

It follows from condition (20.10) that p may be chosen so large that $\| Kh - K_p h \|$ becomes smaller than an arbitrarily small $\epsilon > 0$ for all functions of the family H. One may easily convince onself of this since, by the Cauchy-Schwarz inequality,

$$\left\{ \int [K(P, Q) - K_p(P, Q)] h(Q)\, dQ \right\}^2$$
$$\leq \int [K(P, Q) - K_p(P, Q)]^2\, dQ \int h^2(Q)\, dQ,$$

and

$$\| Kh - K_p h \| \leq M \left[\iint (K(P, Q) - K_p(P, Q))^2\, dPdQ \right]^{\frac{1}{2}} \leq M/2^p.$$

Having chosen p in this way, we can then determine a number N such that, for $n > N$ and $m > N$,

$$\| K_p h_m^{(m)} - K_p h_n^{(n)} \| < \epsilon,$$

since the sequence (20.17) converges in the mean. The left side of (20.18) is hence less than 3ϵ. It follows that Σ is compact.

4. The proof of the existence of finite eigenvalues of the integral equation with symmetric kernel $K(P, Q)$ that is square summable for all of GG, with G bounded, proceeds just as in subsection 3, §12. The difference consists only in the following.

a) The function $\varphi_A(P)$ must be defined so that it vanishes everywhere with the exception of the intersection of the set G with a certain cube K_A, whose mid-point is A and whose sides are parallel to the axes. On this intersection $\varphi_A(P)$ is to be equal to 1. The assumption $\mu_m = \mu_M = 0$ will then lead to the conclusion that the integral

$$\iint K(P, Q)\, dP\, dQ$$

vanishes when taken over the intersection of GG with an arbitrary $2d$-dimensional cube in (P, Q)-space with sides parallel to the axes. Making use of subsection 1 b) of this section, it follows that $K(P, Q)$ must vanish almost everywhere on the topological product of G with itself (true even if G is not a region).

b) We understand by convergence, convergence in the mean, and accordingly shall make use of the theorem proved in the previous subsection of this section instead of the theorem given in subsection 2, §12.

c) There are only finitely many linearly independent eigenvectors to each eigenvalue, and the eigenvalues cannot have a finite limit point. This is shown as follows. As in §13 construct the sequences (13.6) and (13.7) for the kernel $K(P, Q)$. We do not exclude a priori that infinitely many terms of (13.7) are equal. For an arbitrary m we have

$$\iint \left[K(P, Q) - \sum_{i=1}^{m} \varphi_i(P)\varphi_i(Q)/\lambda_i \right]^2 dP\, dQ$$
$$= \iint K^2(P, Q)\, dP\, dQ - \sum_{i=1}^{m} 1/\lambda_i^2 \geq 0.$$

The convergence of the series

$$\sum_{i=1}^{\infty} 1/\lambda_i^2$$

follows and hence $\lim_{i\to\infty} \lambda_i = \infty$.

REMARK. In this way we have shown that an integral equation with a symmetric square summable kernel has only a denumerable set of eigenvalues and linearly independent eigenfunctions. This fact is only a special case of the fact that any orthonormal system S of functions given in a bounded region has a countable cardinal. It follows from Weierstrass's theorem, that to any function $f \in S$ one may assign a polynomial φ_f in the coordinates with rational coefficients so that the norm of the difference $f - \varphi_f$ is smaller than an arbitrarily prescribed ϵ, and in particular, less than $2^{\frac{1}{2}}/2$. But the set of such polynomials is countable. If S were not countable, then one could find functions $f \neq g$ for which $\varphi_f = \varphi_g$. By the triangle inequality (subsection 5, §11) the norm of the difference is less than $2^{\frac{1}{2}}$. But that is not possible since one can easily show that the norm of the difference of two orthogonal normed functions equals $2^{\frac{1}{2}}$.

5. The considerations of §§13–16 remain valid here provided the set G is bounded. We need only take convergence in the mean everywhere instead of uniform convergence. The series (14.3), e.g., converges in the mean by the Hilbert-Schmidt theorem to any function $f(P)$ in the range of the operator.

6. One can prove all three Fredholm theorems for an integral equation (14.7) with a symmetric kernel $K(P, Q)$ of the type under consideration, provided the set G is bounded.

For every function $f(P)$ with summable square appearing on the right side of equation (14.7) one can compute the Fourier coefficients with respect to the orthonormal systems of eigenfunctions (13.6) of the kernel $K(P, Q)$. Due to the Bessel inequality the sum of the squares of these coefficients converges. It follows from the Riesz-Fischer theorem that the series on the right side of (14.10) converges in the mean to a function $\varphi_0(P)$, square summable for every function $f(P)$ that is square summable if λ is not equal to an eigenvalue λ_i. The function $\varphi_0(P)$ satisfies (14.7) almost everywhere. In order to convince ourselves of this fact we first note the following. If the functions $F_1(P)$ and $F_2(P)$ obtained respectively upon substituting $\varphi_0(P)$ in the left and right sides of equation (14.7) were not equal almost everywhere, then the integral of the square of their difference could not be equal to zero. But this is not possible. Substitute for $\varphi(P)$ in the left and right sides of equation (14.7) the expression

$$\lambda \sum_{i=1}^{n} f_i \varphi_i(P)/(\lambda_i - \lambda) + f(P).$$

Let the results of the substitutions be denoted by $F_1^{(n)}(P)$ and $F_2^{(n)}(P)$, respectively. Clearly, for n sufficiently large, the norm (mean square deviation) of the differences $F_1 - F_1^{(n)}$ and $F_2 - F_2^{(n)}$, which we write as

$$\| F_1 - F_1^{(n)} \|, \qquad \| F_2 - F_2^{(n)} \|,$$

will become arbitrarily small. It will follow from the triangle inequality that the relation

$$\lim_{n \to \infty} \| F_1^{(n)} - F_2^{(n)} \| = 0$$

cannot hold if

$$\| F_1 - F_2 \| \neq 0.$$

On the other hand, we have

$$F_2^{(n)}(P) = \lambda^2 \int K(P, Q) \sum_{i=1}^{n} [f_i\varphi_i(Q)/(\lambda_i - \lambda)] \, dQ$$

$$+ \lambda \int K(P, Q)f(Q) \, dQ + f(P).$$

If we use the Hilbert-Schmidt theorem in its new formulation and the fact that the functions φ_i are the eigenfunctions of the kernel $K(P, Q)$, we obtain, almost everywhere,

$$F_2^{(n)}(P) = \sum_{i=1}^{n} (\lambda^2/\lambda_i)f_i\varphi_i(P)/(\lambda_i - \lambda) + \sum_{i=1}^{\infty} (\lambda/\lambda_i)f_i\varphi_i(P) + f(P)$$

$$= \lambda\sum_{i=1}^{n} f_i\varphi_i(P)/(\lambda_i - \lambda) + f(P) + \sum_{i=n+1}^{\infty} (\lambda/\lambda_i)f_i\varphi_i(P).$$

Consequently, one has almost everywhere

$$F_2^{(n)}(P) - F_1^{(n)}(P) = \lambda\sum_{i=n+1}^{\infty} f_i\varphi_i(P)/\lambda_i$$

and therefore also

$$\| F_2^{(n)}(P) - F_1^{(n)}(P) \| = \lambda^2 \sum_{i=n+1}^{\infty} f_i^2/\lambda_i^2 \to 0 \quad \text{as } n \to \infty.$$

The assumption that $\varphi_0(P)$ does not satisfy equation (14.7) almost everywhere thus leads to a contradiction.

In this way we have shown that equation (14.7) has a solution for arbitrary $f(P)$, if λ is not an eigenvalue of this equation. And this is also the only solution. If, namely, $\varphi_1(P)$ and $\varphi_2(P)$ were two solutions of equation (14.7), then $\varphi_1(P) - \varphi_2(P)$ would be a solution of the corresponding homogeneous equation and λ would be an eigenvalue, contrary to assumption. This proves the first Fredholm theorem.

Because of the symmetry of the kernel $K(P, Q)$, the second Fredholm theorem is an evident consequence of Remark c) to subsection 4.

If λ coincides with one of the λ_i, a necessary condition for the existence of a solution of equation (14.7) is that $f(P)$ be orthogonal to all the eigenfunctions $\varphi_1^{(i)}(P), \cdots, \varphi_m^{(i)}(P)$ that belong to this λ_i. One has

$$\int \varphi(P)\varphi_k^{(i)}(P) \, dP = \lambda_i \iint K(P, Q)\varphi(Q)\varphi_k^{(i)}(P) \, dP \, dQ + \int f(P)\varphi_k^{(i)}(P) \, dP$$

$$= \int \varphi(Q)\varphi_k^{(i)}(Q) \, dQ + \int f(P)\varphi_k^{(i)}(P) \, dP, \quad k = 1, 2, \cdots, m.$$

This condition is also sufficient. Form the series (14.10) where we put an arbitrary but fixed constant α_i for $\lambda_i/(\lambda - \lambda_i)$ in those terms where $\lambda = \lambda_i$ (and hence $f_i = 0$). We then show in analogy with the above that the series so obtained converges in the mean and its sum satisfies (14.7) almost everywhere. By varying the α_i, we obtain all solutions of equation (14.7) (which are given by adding all solutions of the homogeneous equation to an arbitrary solution of the non-homogeneous equation). This proves the third Fredholm theorem.

REFERENCES

1. BURKILL, J. C. *The Lebesgue Integral*, Cambridge University Press (Cambridge Tract No. 40), 1953.
2. TITCHMARSH (see Ref. 2, Ch. I), p. 397, Exercise 17.

LIST OF THEOREMS

Alternatives theorems (for a system of algebraic equations)	8
Fredholm theorems—statement	9–10
Existence and uniqueness theorem for kernels of small bound	18, 21
Fredholm theorems for equations with almost degenerate kernels	25
Fredholm theorems for equations with uniformly continuous kernels	28
Fredholm theorems for equations with kernels of the form $\bar{K}(P, Q)/PQ^\alpha$	34
Fredholm theorems for the Fredholm equation ($K = \bar{K}/PQ^\alpha$ on a surface	36
Volterra equations have no eigenvalues	40
Cauchy inequality	46
Cauchy-Schwarz inequality	47
Triangle inequality	46
Fourier coefficients yield least mean square error	48
Bessel's inequality	48
Complete systems are closed	49
Generalized Cauchy-Schwarz inequality	55
The image of a uniformly bounded family is uniformly bounded and equicontinuous	56
The existence of an eigenvalue	57
Orthogonality of eigenfunctions	61
Reality of eigenvalues	62
Deletion theorem	64
Characterization of degenerate kernels	66
Hilbert-Schmidt theorem	68
The Schmidt formula	71
Expansion of the kernel	72
Dini's theorem	74
Uniform convergence of the series for $K_2(Q, Q)$	75
Uniform convergence of the series for $K_2(P, Q)$	75
The series of reciprocal squares of λ_i converges	76
Mercer's theorem	76
Reduction of a quadratic form to canonical form	80
Fubini's theorem	85
Riesz-Fischer theorem	86
Closure implies completeness	86
The existence of an eigenvalue	88
The image of a bounded set of functions by a square integrable kernel is compact	88–89
An eigenvalue possesses only finitely many eigenfunctions (square summable case)	91
Hilbert-Schmidt theorem (square summable case)	92
Fredholm theorems for square summable kernel	92

INDEX

Alternatives theorems (for a system of algebraic equations), 8
Bessel's inequality, 48
Cauchy inequality, 46
Cauchy-Schwarz inequality, 47
— — —, generalized, 55
closed orthonormal systems, 49
completeness of function spaces, 86
— of orthonormal systems, 49
convergence in the mean, 45
convolution, 17
cosine of angle between functions, 47

Eigenfunctions, eigenvalues, 27
equicontinuity, 56

Fourier coefficients of a function, 48
Fredholm equations, 1
— theorems, 9 ff.

Image of a function by kernel, 53
integral equations, definition of, 1
— — of the first kind, 1
— —, Fredholm (second kind), 1
— —, homogeneous, 1
— —, kernel of, 1
— —, linear, 1
— —, resolvent of, 22
— — of the second kind, 1
— —, singular, 38
— —, transposed, 10
— —, Volterra, 40

Kernel, almost degenerate, 24
—, degenerate, 11
—, eigenvalues of, 27
— of the integral equation, 1
—, iterated, 20
—, positive (negative) definite, 74
—, positive (negative) quasidefinite, 74
—, symmetric, 50
—, transpose of, 24

Lebesgue integration, 85
linear dependence of functions, 47

Manifold, 35
maximal orthonormal system, 64

Norm of a function, 44
— of a vector, 44
normed function, 47

Orthogonality of functions, 47
orthogonalization of eigenfunctions, 62
orthonormal system of functions, 49
— — — —, maximal, 64

Parseval's equation, 49
positive definite form, 55, 74
positive (negative) definite kernels, 74
positive (negative) quasidefinite forms and kernels, 74
principle of contraction mappings, 17

Resolvent of an integral equation, 22
resonance of a string, 6

Scalar product of functions, 46
singular integral equation, 38
string, eigenfrequencies of, 5
—, eigenoscillations of, 6
—, eigenvalues of, 5
—, resonance of, 6
summable function, 85
symbolic product, 17

Transpose of an integral equation, 10
— of a kernel, 24
— of a system of algebraic equations, 7
topological product of spaces, 85
triangle inequality, 46

Volterra integral equation, 40